MARINE BIOLOGY

BIOLOGY OF SEMIARID TROPICAL FISH

MARINE BIOLOGY

Additional books in this series can be found on Nova's website
under the Series tab.

Additional e-books in this series can be found on Nova's website
under the e-book tab.

MARINE BIOLOGY

BIOLOGY OF SEMIARID TROPICAL FISH

JOSÉ ROSA GOMES
AND
SATHYABAMA CHELLAPPA
EDITORS

nova publishers
New York

For permission to use material from this book please contact us:
Telephone 631-231-7269; Fax 631-231-8175
Web Site: http://www.novapublishers.com

Additional color graphics may be available in the e-book version of this book.

Library of Congress Cataloging-in-Publication Data

ISBN: 978-1-63117-067-6

Published by Nova Science Publishers, Inc. † New York

CONTENTS

PREFACE

Brazil is home to equatorial, tropical, semi-arid, highland tropical, and subtropical climatic regions. The semiarid region of Northeastern Brazil, considered an area of drought polygon, has distinctive scrub vegetation referred to as Caatinga that covers over 10% of the Brazilian territory with approximately twenty million inhabitants. The Caatinga is considered one of the 37 Wilderness Areas of the World, and it plays an important role in the maintenance of regional macro-ecological processes, as well as indirectly supporting regions with biodiversity and endemism.

The Brazilian semiarid region is also home to Neotropical fish species that occur in diversified aquatic ecosystems, exhibiting different strategies to survive and reproduce. In this scenario of fantastic diversity there exists an intensive and yet insufficiently known interaction of environmental factors and fish communities. To fully understand this it is necessary to know the details about how it works.

This volume assembles research articles on different themes of biology of semiarid fish species. The contents of the ten articles show at least a fraction of the biology and adaptations of fish species to survive and complete their reproductive cycle in the aquatic ecosystems of semiarid Caatinga ecoregion. The articles address aspects of ecological distribution of fish communities, fish reproduction and behavior, anatomy and histology of the digestive tract and feeding habits of fish, and impacts of fish culture and of fish parasites. This information is important to fish biologists, ecologists, and to all individuals with an interest in fish species.

We are greatly indebted to all the authors who have contributed articles to this volume. We are appreciative of the Universidade Federal do Rio Grande do Norte, Brazil, for the continuing support given to fish biology research. We are, as well, appreciative of the *Animal Biology Journal* of Nova Science Publishers, USA, and especially to Dr. Jose Rosa Gomes, Chief Editor of *Animal Biology Journal*, for getting this publication in a special volume.

S. Chellappa Ph. D.
November, 2013.

Section I: Aspects of Ecological Distribution of the Fish Communities

In: Biology of Semiarid Tropical Fish
Editors: J. R. Gomes and S. Chellappa
ISBN: 978-1-63117-067-6

Chapter 1

ENDEMIC FISH COMMUNITIES AND ENVIRONMENTAL VARIABLES OF THE PIRANHAS-ASSU HYDROGRAPHIC BASIN IN THE BRAZILIAN CAATINGA ECOREGION

Wallace Silva do Nascimento[1], Andréa Soares de Araújo[1], Liliane de Lima Gurgel[2], Maria Emília Yamamoto[1], Naithirithi T. Chellappa[3], Ricardo Souza Rosa[4] and Sathyabama Chellappa[1]*

[1]Post-Graduate Programme in Psychobiology, Centro de Biociências, UFRN, Campus Universitário, CEP: Rio Grande do Norte, Brazil

[2]Instituto Federal de Educação Ciência e Tecnologia do Rio Grande do Norte (IFRN), Campus Macau, CEP: RN, Brazil

[3]Department of Oceanography and Limnology, Universidade Federal do Rio Grande do Norte, Praia de Mãe Luiza, s/n, Natal, CEP: RN, Brazil

[4]UFPB, Centro de Ciências Exatas e da Natureza, Departamento de Sistemática e Ecologia, Laboratório de Ictiologia. Cidade Universitária, Castelo Branco, CEP: João Pessoa, Paraíba, Brazil

ABSTRACT

The Brazilian Caatinga ecoregion is home to several endemic fauna of the neotropical region. This paper presents the results of fish faunal surveys conducted in the Piranhas-Assu hydrographic basin located in the Caatinga ecoregion, which is an important area for the conservation of fish species. The fish samples captured were distributed in 4 orders (Characiformes, Perciformes, Siluriformes and Synbranchiformes), 11 families (Characidae, Curimatidae, Auchenipteridae, Anostomidae, Prochilodontidae, Erythrinidae, Cichlidae, Sciaenidae, Heptapteridae, Loricariidae, Synbranchidae) and 22

* To whom correspondence should be addressed: E-mail address: chellappa.sathyabama63@gmail.com, wallacesnbio@hotmail.com, andreaaraujo@unifap.br.

species, of which 17 are endemic to the Caatinga ecoregion. The order Characiformes was more representative with 46.35% followed by Perciformes 35.38%, Siluriformes 17.44% and Synbranchiformes 0.5%. *Oreochomis niloticus*, the only exotic species registered was expressive (24.92%) followed by the native species *Leporinus piau* (18.77 %). Urgent measures must be taken to conserve the endemic fish species of the hydrographic basins located in the Caatinga ecoregion.

Keywords: Endemic fish species, Piranhas-Assu hydrographic basin, Brazilian Caatinga ecoregion

INTRODUCTION

Tropical areas contain the majority of the world's freshwater fish biodiversity (Lowe-McConnell, 1987; Kottelat et al., 1993; Kottelat and Whitten, 1996; Queiroz et al., 2006). The freshwater ichthyofauna of the Neotropical region is characterized by high species richness which contributes to almost 31% of fish diversity of the planet (Reis et al., 2003). More than 15% of the Neotropical fish fauna is composed of migratory species (Carolsfeld et al., 2004), and the common reproductive strategy of the migratory fish involves upstream movements toward the spawning grounds at the beginning of the rainy season (Agostinho et al., 2008).

Brazil is an important part of the Neotropical region which holds the largest biodiversity on the planet, accounting for around 13% of the total world biota (Queiroz et al., 2006). Although most of the country is located within the tropical zone, the climate of Brazil varies considerably from the Northern tropical zone to the temperate zones below the Tropic of Capricorn. Thus Brazil exhibits the equatorial, tropical, semi-arid, highland tropical and subtropical climatic regions. Northeastern Brazil has distinctive scrub vegetation "Caatinga", consisting of xerophytic low thorny bushes adapted to the semi-arid evapotransprAtation driven climate, which covers over 10% of the Brazilian territory. It is located between 3°S 45°W and 17°S 35°W, extending across eight states of Brazil: Piauí, Ceará, Rio Grande do Norte, Paraiba, Pernambuco, Alagoas, Sergipe, Bahia, and parts of Minas Gerais (Leal et al., 2003).

The hydrographic network of the Caatinga ecoregion is seasonal and modest when compared to other Brazilian regions. The seasonality of the freshwater ecosystems of this region is due to recurrence of extended drought, irregular rainfall, high temperatures and an elevated rate of water evaporation (Ab' Saber, 1995; Leal et al., 2003). The fish species that occur in the freshwater ecosystems of this ecoregion are the result of evolutionary processes modulated by climatic factors and hydrological cycles, exhibiting a high degree of endemic species. However, there is need for more detailed information on diversity, endemism and distribution of fish fauna of these ecosystems (Rosa, 2004; Rosa et al., 2005; Chellappa et al., 2009).

The reservoir construction on rivers mainly impact the migratory fish species, which belong to orders Characiformes and Siluriformes, by blocking access to the different habitats needed to complete their life cycle. A series of reservoir constructions on a river can exclude migratory species from the basin or even lead to complete extinction if these species which are endemic to the basin. The disappearance of large migratory fishes from some rivers of the Upper Paraná basin in Brazil, where 140 reservoirs have been constructed could be considered as a good example of this problem (Lowe–McConnell, 1987; Agostinho et al., 2005).

Studies relating to endemic fish species of the Caatinga ecoregion is rather limited (Rosa, 2004; Rosa et al., 2005; Chellappa et al., 2009). This study presents the endemic fish species of the Piranhas-Assu hydrographic basin of the Brazilian Caatinga biome. An attempt is made to identify the main environmental variables which were associated with distribution of fish. This information could contribute to a better management of aquatic environments and for the conservation of the endemic freshwater fish species.

MATERIALS AND METHODS

Study Area

The River Piranhas–Assu hydrographic basin is situated between latitudes 4° and 8° S and longitudes 36° and 39° W (Figure 1). It is located in the Caatinga biome and covers a territory of 44,000 Km^2. It originates in the Serra do Bongo, southeast of Paraiba and enters the State of Rio Grande Norte and eventually flows into the Atlantic Ocean. Around 46 medium to small sized reservoirs have been constructed on this hydrographic basin, which together are responsible for a water storage capacity of 10 million m^3.

Figure 1. Location of Piranhas-Assu hydrographic basin, Rio Grande do Norte, Brazil.

Samples were collected from the River Acauã (latitudes 06° 25' 34,3" S and longitudes 036° 36' 08,3 W) and the Marechal Dutra reservoir (latitudes 06° 25' 00,6" S and longitudes 036° 36" 14,6 W), both located in the Piranhas–Assu hydrographic basin. The fish were captured in a downstream area of the River Acauã since in upper reaches both physical and chemical conditions are stressful, with a lower number of fish species showing physiological and ecological capacities to survive in such places (Matthews and Styron-Junior, 1981). Marechal Dutra is an important reservoir for agriculture and inland fishery purposes, has a maximum depth of 25m, mean depth of 8.5m, water storage capacity of 40 x 10^6 m^3 and a theoretical retention time of 1460 days.

Fish Sampling

Fish samples were captured monthly on a diurnal and nocturnal basis during January to December 2009, with the help of artisanal fishermen of the region, using fishing gear which consisted of stationary nets and gillnets of different mesh sizes (9 – 19 cm), cast nets, hooks and traditional fish traps. Random fish subsamples were collected from a larger catch landed by fishermen, numbered and transported to the laboratory on ice. Morphometric measurements and meristic counts were carried out to check the taxonomical status of each species. Total body length of all fishes were measured (± 1 mm) and body mass recorded (± 1 g). Fish samples were registered in the collection of the Museum of the Department of Systematics and Ecology, Federal University of Paraíba, Brazil.

Data Analysis

The constancy of fish species was calculated using the equation of Dajoz Index (Dajoz, 1973). $C = n / N \times 100$, where: C = constant; n = number of times the species was captured; N = total number of collections. Constant species (C> 50%), accessory species (25 <C <50%) and rare species (C <25%) were calculated from the frequency of occurrence of each species in the sample.

The principal component analysis (PCA) was used in ordering the five environmental variables (water temperature, concentration of dissolved oxygen, pH, electrical conductivity and rainfall) in two factorial axes, with an objective to reduce the dimensionality of the data and to describe the relationships between these environmental variables. The PCA was performed from the linear correlation matrix of environmental variables after logarithmic transformation of data. The ordinations were made in the software XLSTAT 7.5.

Environmental Variables

The values of water temperature (°C), pH, concentration of dissolved oxygen (mgL^{-1}) and electrical conductivity of the water (μScm^{-1}) of the sampling sites were registered between 9:00h and 10:00h on a monthly basis, using portable equipment (Multiparameter kit WTW multi 340i). The mean monthly rainfall data of the region was obtained from the

Meteorological Department of EMPARN (Empresa de Pesquisa Agropecuária do Rio Grande do Norte, Brazil).

RESULTS

Diversity of Icthyofaunal Communities

A total of 602 fish samples were collected (115 from the river and 487 from the reservoir) and the samples were pooled for analysis. The fish were distributed in 4 orders (Characiformes, Perciformes, Siluriformes and Synbranchiformes), 11 families (Characidae, Curimatidae, Auchenipteridae, Anostomidae, Prochilodontidae, Erythrinidae, Cichlidae, Sciaenidae, Heptapteridae, Loricariidae, Synbranchidae) and 22 species (Figure 2), of which 17 were endemic and 4 had been introduced from other hydrographic basins of Brazil and one was an exotic species.

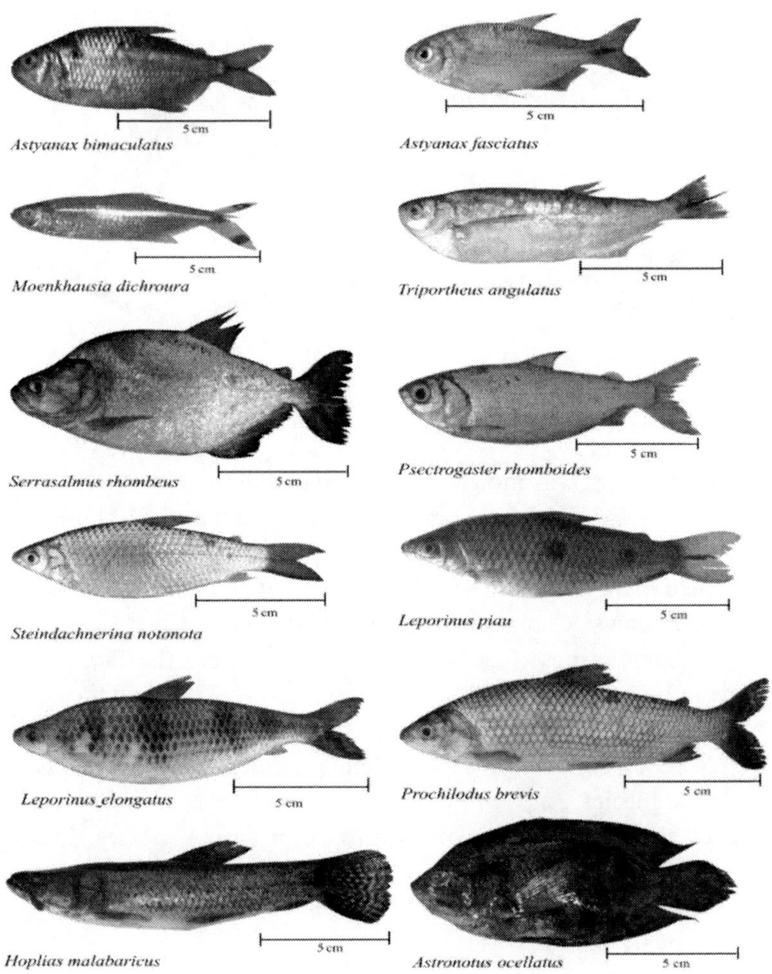

Figure 2. (Continued)

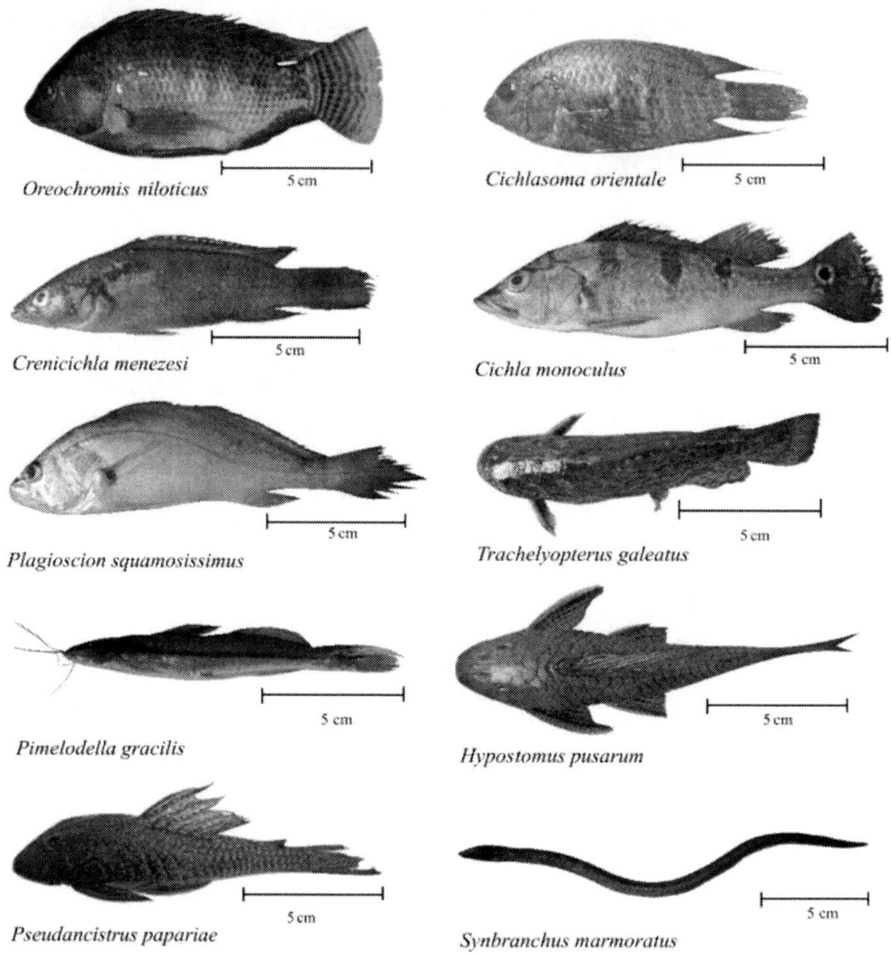

Figure 2. Fish species captured from the Piranhas-Assu hydrographic basin, Brazil.

The order Characiformes was more representative with 46.35% followed by Perciformes 35.38%, Siluriformes 17.44% and Synbranchiformes 0.5%. Tabele 1 shows the fish species, their popular and scientific names, frequency of occurrence and their origin.

The orders which were more representative in number of fish species of the Piranhas-Assu basin were as follows: Characiformes (46.35%), Perciformes (35.38%), Siluriformes (17.44%) and Synbranchiformes (0.50%) (Figure 3a). Among the families, Cichlidae was more representative (29.24%), followed by Anastomidae (18.77%) and the other nine families (Figure 3b). Among the fish species, the exotic fish Nile tilapia, *Oreochomis niloticus* was very expressive (24.92%), followed by the endemic species piau, *Leporinus piau* (18.44 %) and the other endemic species.

Considering the relative frequency of occurrence of the 22 species, 13 were constant, 5 accessory and 4 were rare. Table 2 shows the constancy of occurrence of fish species, maximum, minimum and mean (±SD) values of their total body length (Lt) and body mass (Wt).

Table 1. Fish species captured, their names, absolute frequency (N) and origin in the Piranhas-Assu hydrographis basin, Rio Grande do Norte, Brazil

Order	Characiformes	Common name	N	Origin
	Family Characidae			
	Astyanax bimaculatus (Linnaeus, 1758)	piaba	18	endemic
	Astyanax fasciatus (Cuvier, 1819)	piaba	4	endemic
	Moenkhausia dichroura (Kner, 1858)	piaba	1	endemic
	Triportheus angulatus (Spix & Agassiz, 1829)	sardinha	24	endemic
	Serrasalmus rombeus (Linnaeus, 1766)	red-eye piranha	1	endemic
	Family Curimatidae			
	Psectrogaster rhomboides (Eigenmann & Eigenmann, 1889)	branquinha	25	endemic
	Steindachnerina notonota (Miranda Ribeiro, 1937)	sagüiru	3	endemic
	Family Anostomidae			
	Leporinus piau (Fowler, 1941)	piau	111	endemic
	Leporinus elongatus	piau verdadeiro	2	introduced
Order	Characiformes	Common name	N	Origin
	Family Prochilodontidae			
	Prochilodus brevis (Steindachner, 1875)	curimatã	88	endemic
	Family Erythrinidae			
	Hoplias malabaricus (Bloch, 1794)	traíra	4	endemic
Order	Perciformes			
	Family Cichlidae			
	Astronotus ocellatus (Agassiz, 1831)	Oscar-apaiarí	1	introduced
	Oreochromis niloticus (Linnaeus, 1758)	tilápia	150	exotic
	Cichlasoma orientale (Kullander, 1983)	cará	8	endemic
	Crenicichla menezesi (Ploeg, 1991)	jacundá	11	endemic
	Cichla monoculus (Spix & Agassiz, 1831)	tucunaré	6	introduced
	Family Sciaenidae			
	Plagioscion squamosissimus (Heckel, 1840)	pescada branca	37	introduced
Order	Siluriformes			
	Family Auchenipteridae			
	Trachelyopterus galeatus (Linnaeus, 1766)	cangati	3	endemic
	Family Heptapteridae			
	Pimelodella gracilis (Valenciennes, 1835)	niquim	75	endemic
	Family Loricariidae			
	Pseudancistrus papariae (Fowler, 1941)	cascudo barbudo	17	endemic
	Hypostomus pusarum (Starks, 1913)	cascudo	10	endemic
Order	Synbranchiformes			
	Family Synbranchidae			
	Synbranchus marmoratus (Bloch, 1795)	mussum	3	endemic
TOTAL			602	

Environmental Variables

The period from January to June is considered the traditional rainy season of this region, with the dry period occurring between July and December. During January, 2009 a rainfall of 42.2 mm was registered marking the beginning of the rainy season.

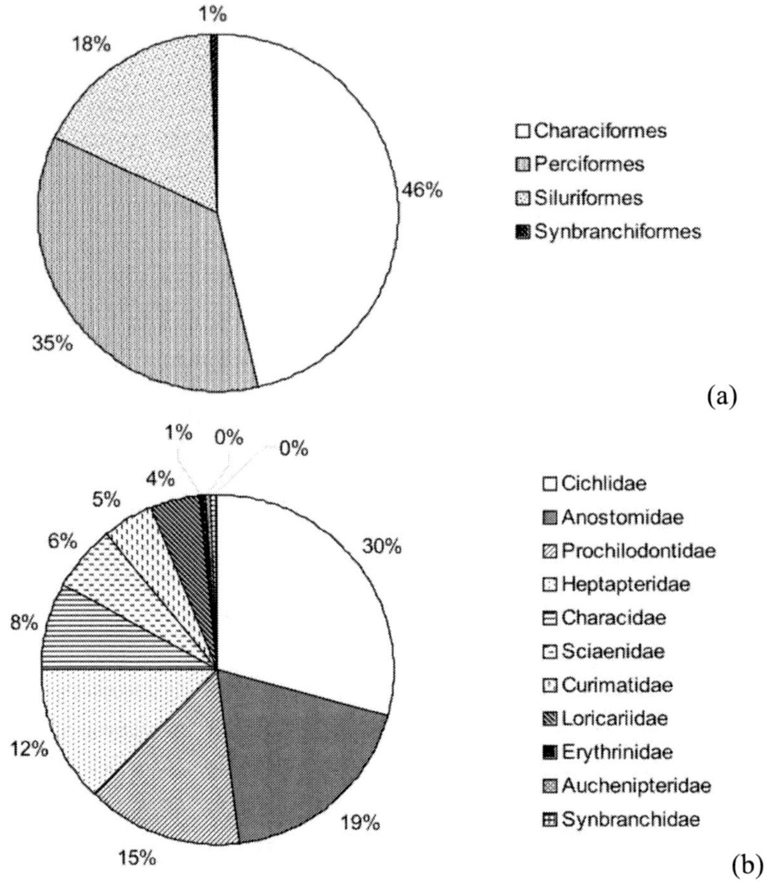

Figure 3. Relative frequency of (a) orders, (b) families of fish species captured from the Piranhas-Assu hydrographic basin, Brazil.

The highest rainfall of 232.2 mm was registered in May and the lowest of 14.6 mm was in June with a mean rainfall of 62.5 mm (± 87.1 SD). No rainfall occurred during August to December, 2009.

From January to December, 2009 the water temperature varied from 25°C to 30.3°C, with a mean of 27 °C (± 1.6 SD). The lowest values were recorded in April (25.5°C) and May (25°C), which correspond to the months of highest rainfall. The highest temperature was recorded in December (30.3°C) during the dry season. Concentration of dissolved oxygen ranged from a minimum of 5.7 mgL^{-1} in August to a maximum of 12.4 mgL^{-1} in February, with a mean of 7.5 mgL^{-1} (± 1.8). Hydrogenionic potential (pH) was alkaline throughout the study period, with a maximum value of 9.4 in November and a minimum of 7.8 in May, with a mean of 8.6 (± 0.5). Mean electrical conductivity was 439.5 μScm^{-1} (± 59.9), the lowest value of 330 μScm^{-1} occurred in May and the highest value of 533 μScm^{-1} occurred in January (Figure 4).

Principal component analysis (PCA), applied to the environmental variables showed that the two first axes together explained 95.05% of the total variance in environmental variables. The first factorial axis explains 80.87% of data variance, while the second explains 14.18%. The ordination diagram shows that environmental variables such as concentration of dissolved oxygen and precipitation were the most influential factors during the rainy season

(January to June 2009), while pH, water temperature and electrical conductivity were the important factors in the dry season (from July to December 2009) (Figure 5).

Table 2. Total length (Lt in cm) maximum, minimum and mean (±SD), weight (Wt in g) maximum, minimum and mean (±SD) and percentage of occurrence (Ci) of fish species captured in Piranhas-Assu hydrographic basin, Rio Grande do Norte, Brazil

Species	Lt Min	Lt Max	Lt Mean ±SD	Wt Min	Wt Max	Wt Mean ±SD	Ci (%)
Astyanax bimaculatus	6.6	7.5	7.1 ±0.4	5.2	7.0	5.6±0.6	Constant (100)
Triportheus signatus	12	18	14.5±0.8	24	45.5	27.5±4.9	Constant (100)
Psectrogaster rhomboides	13.5	15.5	14.73±8.5	30	59.5	42.0±7.9	Constant (100)
Leporinus piau	9,6	30	14,92±4,8	11,5	428	52,9±30,7	Constant (100)
Prochilodus brevis	18	24.5	17.6±8.1	83	210	127.7±115	Constant (100)
Oreochromis niloticus	19	30	17.5±7.1	132	520	162±14.5	Constant (100)
Plagioscion squamosissimus	14.5	34.5	24.5±5.5	74	525	195.3±135	Constant (100)
Pimelodella gracillis	8.5	16.5	13±1.8	2.5	26	14.1±5.8	Constant (100)
Cichlasoma orientale	12.3	15	13.7±1.2	44	78	61.1±10	Constant (60)
Pseudancistrus papariae	18	21.5	19.5±2.1	95	205	130.9±53.5	Constant (60)
Hypostomus pusarum	17.5	22.5	9.1±2.4	91	203	129.7±52	Constant (60)
Crenicichla menezesi	9.5	18.5	13.94±1.9	8.5	69.5	32.7±13.2	Constant (80)
Cichla monoculus	13.5	27.5	21.1±5.5	65	310	166.6±110.2	Constant (80)
Steindachnerina notonota	12	13	12.51±0.5	14	18	15.7±2.1	Accessory (40)
Astyanax fasciatus	7.3	7.6	7.5±0.13	5.9	6	5.9±0.12	Accessory (40)
Hoplias malabaricus	30	45	39.7±4.7	420	1000	825±170.8	Accessory (40)
Trachelyopterus galeatus	15	20	17.7±2.5	68	125	95.8±28.3	Accessory (40)
Synbranchus marmoratus	26.5	42	35.5±8	19	230	146.7±11.5	Accessory (60)
Leporinus elongatus	33	33	-	540	540	-	Rare (20)
Astronotus ocellatus	22.5	22.5	-	248.5	248.5	-	Rare (20)
Moenkhausia dichroura	7.5	7.5	-	4	4	-	Rare (20)
Serrasalmus rhombeus	23	23	-	193.5	193.5	-	Rare (20)

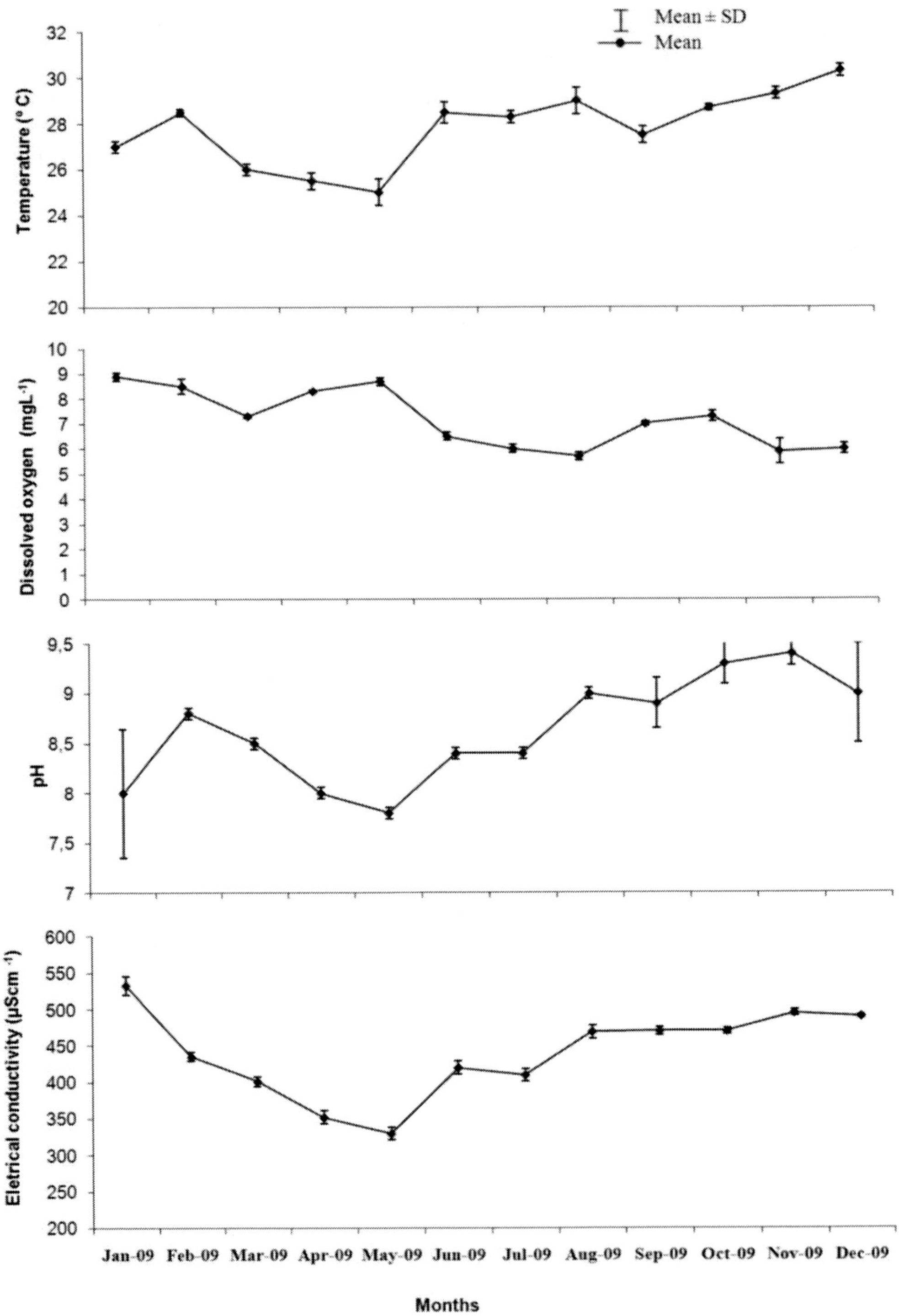

Figure 4. Limnological variables: temperature, concentration of dissolved oxygen, pH and electrical conductivity.

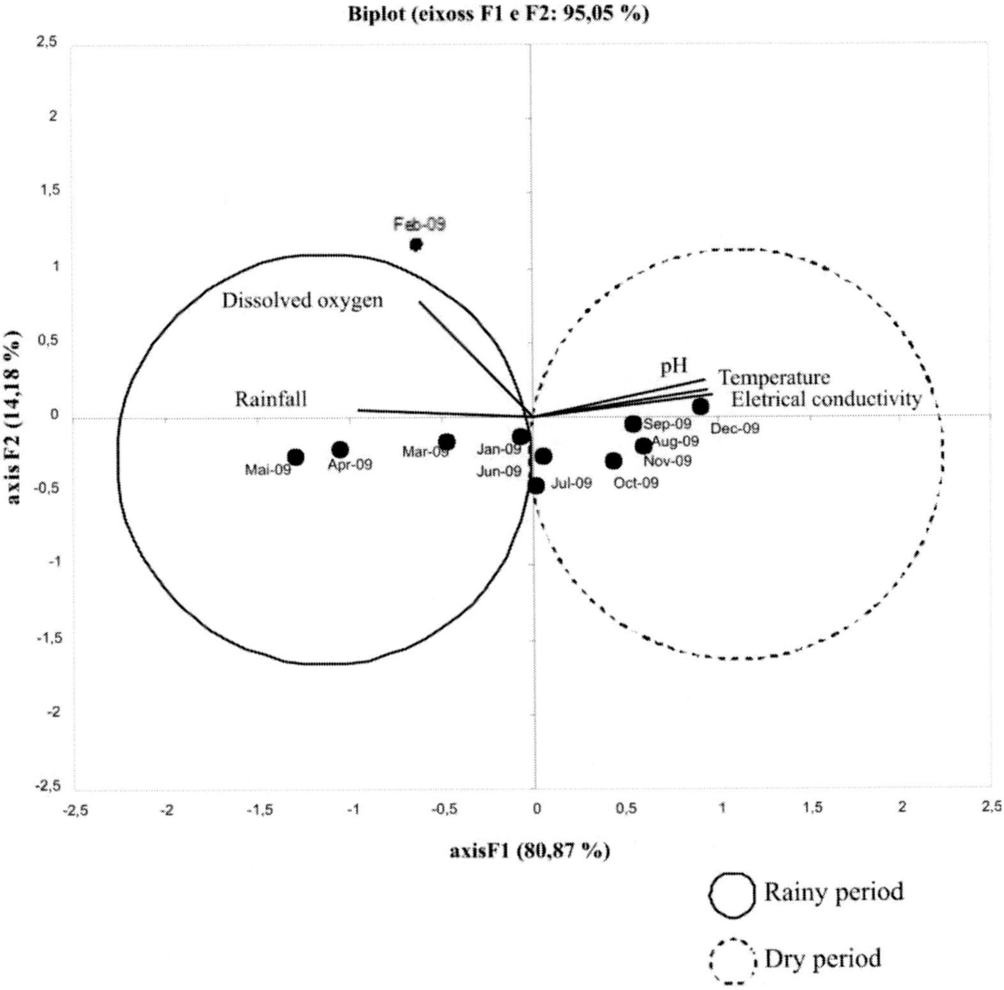

Figure 5. Principal component analysis (PCA) of the variables: rainfall, water temperature, pH, electrical conductivity and dissolved oxygen.

DISCUSSION

Diversity of Icthyofaunal Communities

The 22 fish species recorded in this study contribute significantly to increasing the knowledge of ictiofaunal species in the State of Rio Grande do Norte (RN), Brazil, located in the Caatinga ecoregion, since the previous studies conducted in this State described less than half the number of the species currently registered. Menescal et al. (2000) examined the freshwater fish landings from 1971 to 2000, listing seven fish species of commercial value in the Marechal Dutra reservoir. Chellappa et al. (2009) investigated on the reproductive seasonality and limnoecology of 12 native fish species pertaining to 4 orders and 9 families of this region.

An earlier taxonomical survey of fish fauna in several hydrographic basins of the entire Caatinga ecoregion registered the presence of 240 species, distributed in 7 orders. The order Siluriformes had the highest diversity with 101 species (family Loricariidae with 34 species; Pimelodidae with 27 species and Callichthyidae with 15 species). The order Characiformes presented the second highest diversity, with 89 species (family Characidae with 50 species). Among the Cyprinodontiformes, the family Rivulidae had the highest diversity with 24 espécies. The rest of the orders were represented by Gymnotiformes (6), Perciformes (17), Myliobatiformes (1), Clupeiformes (1) e Symbranchiformes (1). Of the total fish species encountered, 9 were introduced and 136 were considered endemic to Caatinga. The 9 introduced species were: *Poecilia latipinna, P. reticulata, Plagioscion squamosissimus, Astronotus ocellatus, C. monoculus, C. ocellaris, Geophagus surinamensis, O. niloticus* and *Tilápia rendalli* (Rosa *et al.*, 2005).

A study conducted in the freshwater environments of a reservoir on the Sharavathi River in India, described 43 tropical fish species, with a dominance of the family Cyprinidae (Sreekanth and Ramachandra, 2005). Bhakta and Bandyopadhyay (2008) investigated the freshwater perennial ecosystems located in West Bengal, India, and documented 45 fish species, distributed in 8 orders, 18 families and 29 genera. In Nyagui River basin, located in Zimbabwe, Africa, fish diversity was represented by 24 species and 8 families, with a dominance of the family Cyprinidae (Kadye and Marshall, 2007). Kouamélan et al. (2003), recorded 40 fish species distributed in 17 families and 26 genera in the hydrographic basin of the River Boubo in Ivory Coast.

According to Livingstone et al. (1982) large rivers have more fish species than smaller ones and there is a strong correlation between the volume of water discharge, the area of drainage basin and the number of species. These relationships have been explained due to the fact that rivers with large drainage basins provide a greater diversity of habitats (Hugueny, 1989). The river where the sampling was done during this work is small and consequently has fish fauna with a low diversity in comparison to large tropical rivers. Introduction of fish species from other hydrographic basins, and especially the introduction of the exotic Nile tilapia, could have possibly contributed to the reduction of native fish species in River Acauã.

The high frequency of the fish pertaining to the order Characiformes registered in the present study is mainly due to the following reasons: (1) the order Characiformes includes most fish species encountered in the freshwater ecosystems of Brazil; (2) there is predominance of small sized fish species among the Characids, which are able to complete their life cycle in calm waters. Predominance of Characids has also been observed in a temporary Amazon lake (Chellappa et al., 2005), in the Paraná hydrographic basin (Hoffmann et al., 2005; Langeani et al., 2007; Agostinho et al., 2007) and in the Amazonian small streams (Mojica et al., 2009).

The exotic Nile tilapia, *O. niloticus* outstands among the representative species in the Piranhas–Assu hydrographic basin. Tilapia has been introduced in the freshwater ecosystems of the Caatinga ecoregion since the beginning of 1970 in order to upgrade the fish production and fish culture (Gurgel and Fernando, 1994). Presently, it is a dominant species in most reservoirs of northeastern Brazil. This is due to its rapid growth, omnivorous food habits, high reproductive efficiency, parental care and resistance to environmental variations (Zambrano et al., 2006). In 2002, 87% of fish production in Brazil was based on the introduced species (Casal, 2006), and total gross production of tilapia of the genus *Oreochromis* increased exponentially from 1970 to 2006 (Vitule, et al., 2009). Menescal et al. (2000) examined the fish landings in Marechal Dutra reservoir from 1971 to 2000 and observed changes in fish community structure after the introduction of Nile tilapia, with damages caused to the native species, such as, *Prochilodus brevis, Leporinus* sp. and *Hoplias malabaricus*. Besides tilapia other predatory species were introduced in this reservoir, such as: *Plagioscion squamosissimus*, *Astronotus ocellatus* and *Cichla monoculus*. A new predator in freshwater ecosystems often results in a total change in community structure, as it had occurred with the introduction of Nile perch in Lake Victoria in east Africa (Gophen et al., 1995) and the introduction *Cichla ocellaris* in Lake Gatun in Panama (Zarat and Paine, 1973).

The endemic fishes are of great ecological importance. The loss of diversity through the introduction of species and construction of reservoirs on rivers could cause significant damage not only the community but also the ecosystem as a whole. Taylor et al (2006) emphasized on the importance of endemic species to its ecosystem when they investigated as to how the impact of the loss of a dominant migratory detritivorous fish species *P. mariae*, could alter the metabolism of ecosystems and the flow of organic carbon in an Andean river located in the Orinoco basin, which led to the degradation of the whole river.

Diversified factors influence the constant occurrence of fish species. The environmental variables associated with seasonality could change the number of individuals, altering the number of occurrence during a specific season. The occasional species are those which migrate for feeding and reproductive purposes. Some species such as, *M. dichroura, S. rombeus, L. elongatus* and *A. ocellatus* are considered rare.

Ichthyofaunal surveys conducted in the reservoirs of Brazil showed a gradual decrease in the richness and diversity of fish over time, since the formation of the reservoirs (Agostinho et al., 1999; Benedito-Cecílio and Agostinho, 2000; Agostinho, et al., 2005; Agostinho et al., 2008). This may be due to the deleterious effect exerted by the changing conditions of the reservoirs on the fish communities (Smith et al., 1993). The low diversity of fish in the reservoir could be explained by its age (over 30 years), operational procedures of the reservoir and uncontrolled fishing. Species richness may be associated with the number of samples (effort) and the reservoir area (Araujo-Lima et al., 1995), requiring care in comparison to other reservoirs.

In tropical environments the temperature variations are influenced by the rainfall fluctuations. Rainfall stands out among the environmental variables, causing the carriage of soil nutrients, increasing the availability of food and enriching the aquatic ecosystems. Temperature oscillations in accordance with the fluctuations in rainfall show the seasonality of the region. The decrease in the level of available dissolved oxygen during the dry season could be due to utilization of oxygen for respiratory purposes by organisms and decomposition of organic matter that is usually abundant in tropical reservoirs. Aquatic ecosystems with alkaline pH values are found in regions with negative water balance, such as the reservoirs in Northeast Brazil (Chellappa and Costa, 2003).

The values of electrical conductivity remained low throughout the study period, with increase during prolonged periods of drought and consequent dilution during the rainy season (Chellappa et al., 2003). The relationship between environmental variables shows that rainfall influences these variables, showing its importance in the seasonality of the Caatinga region. The abundance of fish was significantly related to the environmental variables. Matthews (1998), Matthews et al. (1992) and Agostinho et al. (2004) verified the importance of limnological variables in structuring fish communities in different types of freshwater environments.

The conservation of fish diversity is a major global environmental challenge (Park et al., 2003). For this, serious and urgent political action is needed to curb human activities, such as over fishing; release of chemicals, agricultural waste, urban and industrial sewage effluents in aquatic ecosystems; the construction of reservoirs, dams and canals; the extraction of sand and other minerals; destruction of riparian vegetation and wetland fringe, and the introduction of alien or exotic fish species. These factors are the main causes of loss of fish diversity. There are few management programs to conserve the diversity of freshwater fish, which are intended to fisheries policies, such as the identification and listing of species, the determination of population size and evaluation of fish reproductive behavior. These measures are of fundamental importance for the conservation of fish.

ACKNOWLEDGMENTS

This study was funded by the National Council for Scientific and Technological Development of Brazil (CNPq). Thanks are due to CAPES for the financial support awarded during the study period to the second and third authors.

REFERENCES

Ab′ Saber, A.N. 1995. *The Caatinga Domain*. In: Monteiro, S. and Kaz, L., eds. Caatinga-Setão, Sertanejos. Rio de Janeiro, Editora Livroarte, pp. 47-55.

Agostinho, A.A. Miranda, LE., Bini, LM., Gomes, LC., Thomaz, SM. and Suzuki, HI. 1999. Patterns of colonization in neotropical reservoirs, and prognoses on aging. In: Tundisi, J.G.; Straskraba. M., eds. *Theoretical Reservoir Ecology and its Applications*. São Carlos: IIE; AH Leiden, Dordrecht: Backhuys Publishers, pp. 227- 265.

Agostinho, A.A., Gomes, L.C. and Latini, J.D. 2004. Fisheries management in Brazilian reservoirs: lessons from/for South America. *Interciencia* 29**:** 334–338.

Agostinho, A.A., Pelicice, F.M. and Gomes, L.C. 2008. Dams and the fish fauna of the Neotropical region: impacts and management related to diversity and fisheries. *Brazilian Journal of Biology* 68: 1119-1132.

Agostinho, A.A., Thomaz, S.M. and Gomes, L.C. 2005. Conservation of the biodiversity of Brazil's inland waters. *Conservation Biology* 19: 646-652.

Agostinho, AA., Pelicice, FM., Petry, AC., Gomes, LC. and Júlio Jr, HF. 2007. Fish diversity in the upper Paraná River basin: habitats, fisheries, management and conservation. *Journal of Aquatic Ecosystems* 10: 174-186.

Araujo-Lima, C.A.R.M. Agostinho, A.A. and Fabré, N.N. 1995. Trophic aspects of fish communities in brazilian rivers and reservoirs. *Limnology in Brazil*. Rio de Janeiro: ABC/SBL. pp. 105-136.

Benedito-Cecilio, E. and Agostinho, A.A. 2000. Distribution, abundance and use of different environments by dominant ichthyofauna in the influence area of Itaipu Reservoir. *Acta Scientiarum,* 22:429-437.

Bhakta, J. N. and Bandyopadhyay, P. K. 2008. Fish Diversity in Freshwater Perennial Water Bodies in East Midnapore District of West Bengal, India. *International Journal on Environmental Research* 2: 255-260.

Carolsfeld, J., Harvey, B., Ross, C. and Baer, A. 2004. Migratory fishes of South America: biology, fisheries and conservation status. *World Fisheries Trust*, World Bank, Ottawa: IDRC. 380p.

Casal, C.M.V. 2006. Global documentation of fish introductions: the growing crisis and recommendations for action. *Biological Invasions* 8: 3–11.

Chellappa, N.T. and Costa, M.A.M. 2003. Dominant and co-existing species of Cyanobacteria from a eutrophicated reservoir of Rio Grande Norte State, Brazil. *Acta Oecologia* 24: 3–10.

Chellappa, S. Câmara, M.R. and Chellappa, N.T. 2003. Ecology of *Cichla monoculus* (Osteichthyes: Cichlidae) from a reservoir in the semi-arid region of Brazil. *Hydrobiologia 504*: 267-273.

Chellappa, S., Bueno, R.M.X., Chellappa, T., Chellappa, N.T. and Val, V.M.F.A. 2009. Reproductive seasonality of the fish fauna and limnoecology of semi-arid Brazilian reservoirs. *Limnologica* 39: 325-329.

Chellappa, S., Oliveira, J.C.S. and Chellappa, N.T. 2005. Fish fauna of a temporary lake in an Amazonian Conservation Area. *Acta Limnologica Brasiliensia* 17: 283-289.

Dajoz, R. *Ecologia geral*. Petrópolis: Editora Vozes, 1973. 471p.

Gophen, M., Ochumba, P.B.O. and Kaufman, L.S. 1995. Some aspects of perturbation in the structure and biodiversity of the ecosystem of Lake Victoria (East Africa). *Aquatic Living Resources* 8:27-41.

Gurgel, J.J.S. and Fernando, C.H. 1994. Fisheries in semiarid northeast Brazil with special reference to the role of tilapias. *Hydrobiologia* 79: 77–94.

Hoffmann, A. C., Orsi, M. L. and Shibatta, O. 2005. A. Fish diversity in the UHE Escola Engenharia Mackenzie (Capivara) reservoir, Paranapanema River, upper Rio Paraná basin, Brazil, and the importance of large tributaries in its maitenance. Iheringia, *Série Zoologia* 95: 319-325.

Hugueny, B. 1989. West African rivers as biogeographic islands: species richness of fish communities. *Oecologia* 79: 236–243.

Kadye, W.T. and Marshall, B.E. 2007. Habitat diversity and fish assemblages in an African river basin (Nyagui River, Zimbabwe). *African Journal of Ecology* 45: 374–381.

Kottelat, M. and A.J. Whitten, 1996. Freshwater fishes of Western Indonesia and Sulawesi: additions and corrections. *Periplus Editions*, Hong Kong. 8p.

Kottelat, M., A.J. Whitten, S.N. Kartikasari and S. Wirjoatmodjo, 1993. Freshwater Fishes of Western Indonesia and Sulawesi. *Periplus Editions* Ltd., Hong Kong. 221p.

Kouamélan, E. P., Teugels, G.G., N'Douba, V., Gooré Bi, G. and Koné, T. 2003. Fish diversity and it s relationships with environmental variables in a West African basin. *Hydrobiologia* 505: 139–146.

Leal, I.R., M. Tabarelli and J.M.C. Silva. 2003. Ecologia e conservação da Caatinga. Editora Universitária, Universidade Federal de Pernambuco, Recife, Brasil. 822p.

Livingstone, D.A., Rowland, M. and Bailey, P.E. 1982. On the size of African riverine fish faunas. *American Zoologist* 22: 361–369.

Lowe-McConnell, R.H., 1987. *Ecological Studies in Tropical Fish Communities*. Cambridge University Press, Cambridge. 382p.

Matthews, W.J. 1998. *Patterns in Freshwater Fish Ecology,* Chapman and Hall, International Thomson Publishing, New York. 756p.

Matthews, W.J., Hough, J.D. and Robison, W.H. 1992. Similarities in fish distribution and water quality patterns in streams of Arkansas: congruence of multivariate analyses. *Copeia* 2: 296-303.

Matthews, W.J. and Styron-Junior, J.T. 1981. Tolerance of headwater vs. mainstream fishes for abrupt physicochemical changes. *American Midland Naturalist* 105: 149-158.

Menescal, R.A., Oliveira, J.C.S., Campos, C.E.C., Araújo, A.S. and Freire, A.G., 2000. Fish production in Marechal Dutra Reservoir, Acari, RN. *Revista de Ecologia Aquática Tropical* 10: 135–139.

Mojica, J.I., Castellanos, C. and Lobón-Cerviá, J. 2009. High temporal species turnover enhances the complexity of fish assemblages in Amazonian Terra firme streams. *Ecology of Freshwater Fish 18*: 520-526.

Park, Y.S., Chang, J., Lek, S., Cao, W. and Brosse, S. 2003. Conservation strategies for endemic fish species threatened by the Three Gorges Dam. *Conservation Biology* 17: 1748-1758.

Queiroz, L.P., Rapini, A. and Giulietti, A.M. 2006. *Towards greater knowledge of the Brazilian Semi-arid Biodiversity.* Ministerio da Ciência e Tecnologia, Brasília. 142p.

Reis, R. E., S. O. Kullander and C. J. Ferrari Jr. 2003. *Check list of the freshwater fishes of South and Central America*. EDPUCRS: Porto Alegre, Brazil.

Rosa, R.S. 2004. Diversidade e conservação dos peixes da Caatinga. In: Silva, J.M.C., Tabarelli, M., Fonseca, M.T. and Lins, L.V. orgs. *Biodiversidade da Caatinga: áreas e ações prioritárias para a conservação*. MMA, Brasília, pp. 149–161.

Rosa, R.S., Menezes, N.A., Britski, H.A., Costa, W.J.E.M. and Groth, F., 2005. Diversidade, padrões de distribuição e conservação dos peixes da Caatinga. In: Leal, I.R., Tabarelli, M. and Silva, J.M.C. eds. *Ecologia e Conservação da Caatinga*. Editora UFPE, Recife, pp. 135–180.

Smith, E.P., Orvos, R. D. and Cairns Jr, J.1993. Impact assessment using the before- after-control-impact (BACI) model: concerns and comments. *Canadian Journal of Fisheries and Aquatic Resources* 50: 627- 637.

Sreekanth, A. and Ramachandra, T.V. 2005. Fish diversity in Linganamakki Reservoir Sharavathi River. *Ecology, Environment and Conservation* 11: 337-348.

Taylor, B.W., Flecker, A.S. and Hall Jr., R.O. 2006. Loss of a harvested fish species disrupts carbon flow in a diverse tropical river. *Science* 313: 833–836.

Ter Braak, C.J.F. and P. Smilauer. 2002. CANOCO Reference manual and Cano Draw for Windows user's guide: Softwarefor Canonical Community Ordination (version 4.5). Ithaca, Microcomputer Power, 500p.

Vitule, S. R. J., Freire, A. C. and Simberloff, D. 2009. Introduction of non-native freshwater fish can certainly be bad. *Fish and Fisheries* 10: 98–108.

Zambrano, L., Martinez-Meyer, E., Menezes, N.A. and Petersen, A.T. 2006. Invasive potential of common carp (*Cyprinus carpio*) and Nile tilapia (*Oreochromis niloticus*) in American freshwater systems. *Canadian Journal of Fisheries and Aquatic Sciences* 63: 1906– 1910.

Zarat, T.M. and Paine, R.T. 1973. Species introduction in a tropical lake. *Science* 182: 449 – 455.

Section II: Aspects of the Fish Reproduction and Behaviour

In: Biology of Semiarid Tropical Fish
Editors: J. R. Gomes and S. Chellappa
ISBN: 978-1-63117-067-6

Chapter 2

OVARIAN DEVELOPMENT AND REPRODUCTIVE PERIOD OF WHITE MULLET, *MUGIL CUREMA* IN THE COASTAL WATERS OF NORTHEASTERN BRAZIL

Mônica Rocha de Oliveira*[1], *Eudriano Florêncio dos Santos Costa*[2] *and Sathyabama Chellappa*[1]
[1]Post-Graduate Programme in Psychobiology, Department of Physiology, Center of Bioscience, Universidade Federal do Rio Grande do Norte (UFRN), Campus Universitário, Rio Grande do Norte, Brazil
[2]Postgraduate Programme in Oceanography, Institute of Oceanography, Universidade de São Paulo, São Paulo, SP. Brazil

ABSTRACT

White mullet, *Mugil curema* is a coastal pelagic fish and is an important fishery resource of the Atlantic Ocean and is a component of artisanal fisheries in northeastern Brazil. This study reports on the ovarian development and reproductive period of *M. curema* (Osteichthyes: Mugilidae) using both macroscopic and histological techniques. The length-weight relationship, size at sexual maturity and gonadosomatic index (GSI) of females were determined. The females show isometric growth and the size at first sexual maturity was at 24.3 cm of total length. The macroscopic characteristics of the ovaries revealed four stages of development: immature, in maturation, mature and spent and the microscopic features showed five stages: immature, initial maturation, final maturation, mature and spent. *M. curema* has a prolonged reproductive period, with two peaks of reproductive activity in January-February and in May, coinciding with increased rainfall.

Keywords: Ovarian development, reproductive period, size at first sexual maturity, oocyte development

* E-mail: monicaufrn@yahoo.com.br; eudriano@yahoo.com.br; chellappa.sathyabama63@gmail.com.

Introduction

Mullets belonging to the family Mugilidae are distributed in coastal waters, estuaries and freshwater environments in tropical and subtropical regions throughout the world (Harrison, 2003; Nelson, 2006). They are considered as important economic resources that support many small communities through both fishing and aquaculture (Pina and Chaves, 2005; Katselis et al., 2005). Among the seven species of mullets which occur in Brazil, *Mugil curema, M. liza* and *M. platanus* are commercially exploited (Menezes, 1983). White mullet, *M. curema* is a coastal pelagic fish which occurs in the Western Atlantic (Nova Scotia to Argentina), Eastern Atlantic (Senegal to Namibia), and in the Eastern Pacific (California to Chile) (Harrison, 2002). It is an important component of artisanal fishery and is a source of income and food for people depending on fishing in the coastal region of Brazil.

Conservation of fish stocks in their natural habitat are usually endangered by abusive fishing of immature fishes which have not yet completed their reproductive cycle, as recruitment via reproduction is the means by which the resource is renewed. The Fishery Research Program of Northeastern Brazil (REVIZEE-SCORE-NE) has carried out various fishery surveys to determine the conservation status of the of fish stocks in their natural habitats (Lessa, 2006). Fishery surveys carried out in the northeastern continental shelf waters emphasize the importance of the flying fish, *Hirundichthys affinis,* thread herring, *Opisthonema oglinium* and *M. curema* as important fish resource for the State of Rio Grande do Norte, Brazil (IBAMA, 2007).

Reproductive biology of white mullet, *M. curema* was investigated in Margarita Island, Venezuela (Marín et al., 2003); in a coastal lagoon in the Gulf of Mexico (Ibáñez-Aguirre and Gallardo-Cabello, 2004), in the southern Caribbean (Solomon and Ramnarine, 2007). Gonadal development and spawning season of *M. curema* was investigated in southeastern Brazil (Albieri et al., 2009; Albieri et al., 2010). Yet for all, there exists a lacuna of information dealing with ovarian development and reproductive period of *M. curema* in the Northeastern coastal waters of Brazil.

The present investigation aimed to (i) examine the length-weight relationship of female *M. curema*; (ii) determine length at first maturity of females; (iii) estimate fecundity, ovarian developmental stages, type of spawning, relate maturity stages to gonadosomatic index and correlate the reproductive period with rainfall. This information could contribute for the management of this important fishery resource in the coastal waters of Northeastern Brazil.

Materials and Methods

Study Site and Sample Collection

Specimens of *M. curema* were collected monthly during the period August 2008 to July 2009, off Rio Grande do Norte, in the Atlantic coastal waters of Brazil (Figure 1). The specimens were captured using beach seine nets, which were 110 m long, 3 meters in height with a central mesh of 10 mm and 70 mm at the extremities. The beach seine was taken 100 m from the beach by local fishermen and released at a depth of 5 m with the aid of a non-powered local craft.

After the capture, each specimen was measured (total length in cm TL) and weighed (total body mass TW in g). The fish were dissected and the gonads were removed for identification of sex and to assess the stage of development. A total of 180 females collected were separated for detailed studies. Rainfall data of the region was obtained from Agricultural Research Company of Rio Grande do Norte (EMPARN), Brazil.

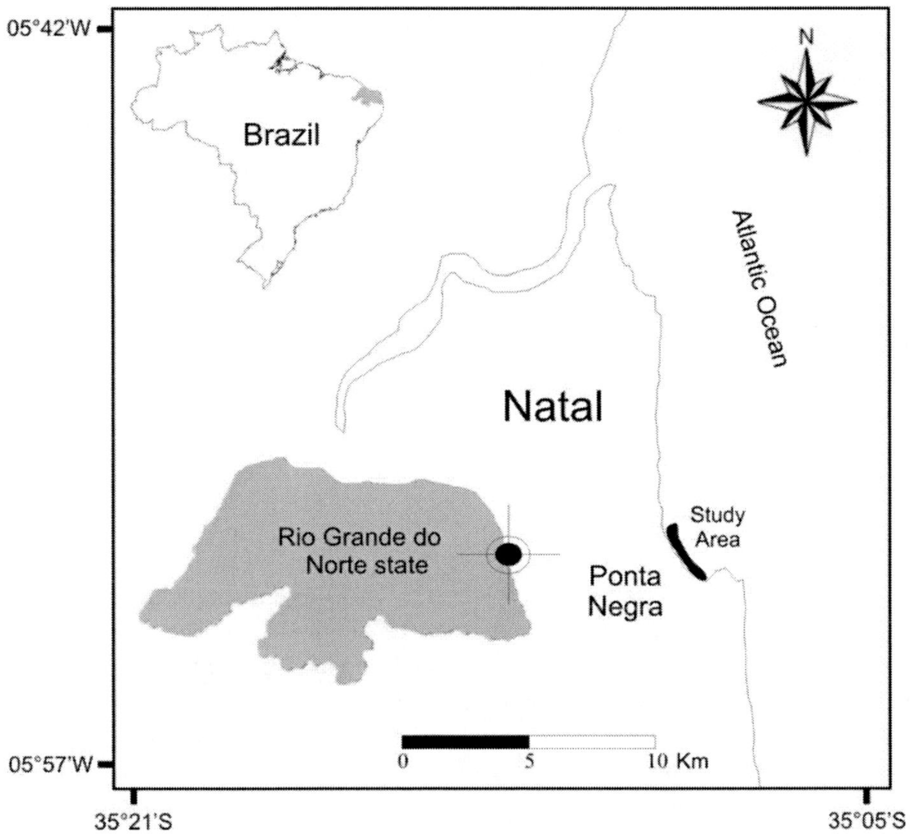

Figure 1. Study area: coastal waters of Rio Grande do Norte, Northeastern Brazil.

Length – Weight Relationship of Females

The length-weight relationship was determined by the equation: $W = a\ L^b$, where W is the total weight (g), L is the total length (cm), a is the interceptor (initial growth coefficient or condition factor) and b is the slope (growth coefficient or fish relative growth rate) (Hayes et al., 1995; Jobling, 2002; Santos et al., 2002). The parameters a and b of L - W relationships were estimated using a power function in which was used the coefficient of determination r^2 to measure the degree of association between two variables (L and W).

Length at First Maturity (L_{50}) of Females

To determine the size at first sexual maturity, the gonads of mature females were pooled (maturing, mature and spawned), followed by distribution of the accumulated relative frequencies by total length classes (Moreno et al., 2005).

Macroscopic Description of Ovaries and Histology and Oocyte Development

The characteristics used for macroscopic classification of the ovaries were according to Vazzoler (1996) and Murua et al. (2003). The following external aspects were considered: size, form, colour, presence of blood vessels, oocytes visible to naked eye, rigidity, turgidity and the space occupied in the body cavity.

In order to avoid possible variation in the developmental stage of oocytes due to their position in the ovaries, histological examinations were carried out on sections from the anterior (cephalic), middle (central), and posterior (caudal) regions of 20 ovaries in different developmental stages (Yoshida, 1964). Fragments of ovaries selected for histological analysis were fixed in Bouin solution for 12-24 h (depending on size), washed for 24 hours in running water to remove excess fixative and were later preserved in 70% alcohol. Fragments of ovaries selected for histological study were embedded in paraffin, sectioned at 3-5 μm thickness, and stained with Hematoxylin-Eosin (HE). The histological description of the stages of oocyte development and classification of microscopic stages of ovarian maturation was performed using the terminology described by West (1990) and Vazzoler (1996).

Fecundity and Type of Spawning

The oocytes from mature ovaries were dissociated using Gilson solution and three sub-samples of 0.1 g were extracted and the oocytes were counted using Bogorov plates, a stereo-microscope and an ocular micrometer. Total fecundity was estimated for the total weight of the ovaries. Fecundity = [(number of mature oocytes in the fragments of ovary) x (total weight of ovary)] / (weight of the fragment of the ovary). Absolute fecundity was estimated using mature ovaries weighing on an average 37.96 g of females with average total length of 30.6 cm. The type of spawning was assessed by measuring the oocyte diameter (in μm) and by histological analysis of the oocyte development (Vazzoler, 1996).

Gonadosomatic Index and Reproductive Period

The gonadosomatic index (GSI) of females was calculated using the equation: $GSI = (Wg / Wt) * 100$, where Wg is the gonad weight (g), and Wt is the total weight of the individual (Wootton et al., 1978). The reproductive period was determined based on the distribution of relative frequencies (%) of each maturity stage of gonads and the variation of average monthly gonadosomatic index (GSI).

RESULTS

Sample Size

A total of 366 samples of *M. curema* were captured during the study period, of which 50.8% (n = 186) were males and 49.1% (n = 180) were females. Only the females were used in this study. The frequency distribution of occurrence by total length classes indicate that females ranged from 15.6 to 34.5 cm (24.9 ± 4.1) (Figure 2a). About 35% of females sampled were in the length classes of 19-21 and 21-23 cm and were considered as young females. In terms of body mass, females ranged from 35.4 to 382g (160.5 ± 82.3g) (Figure 2b).

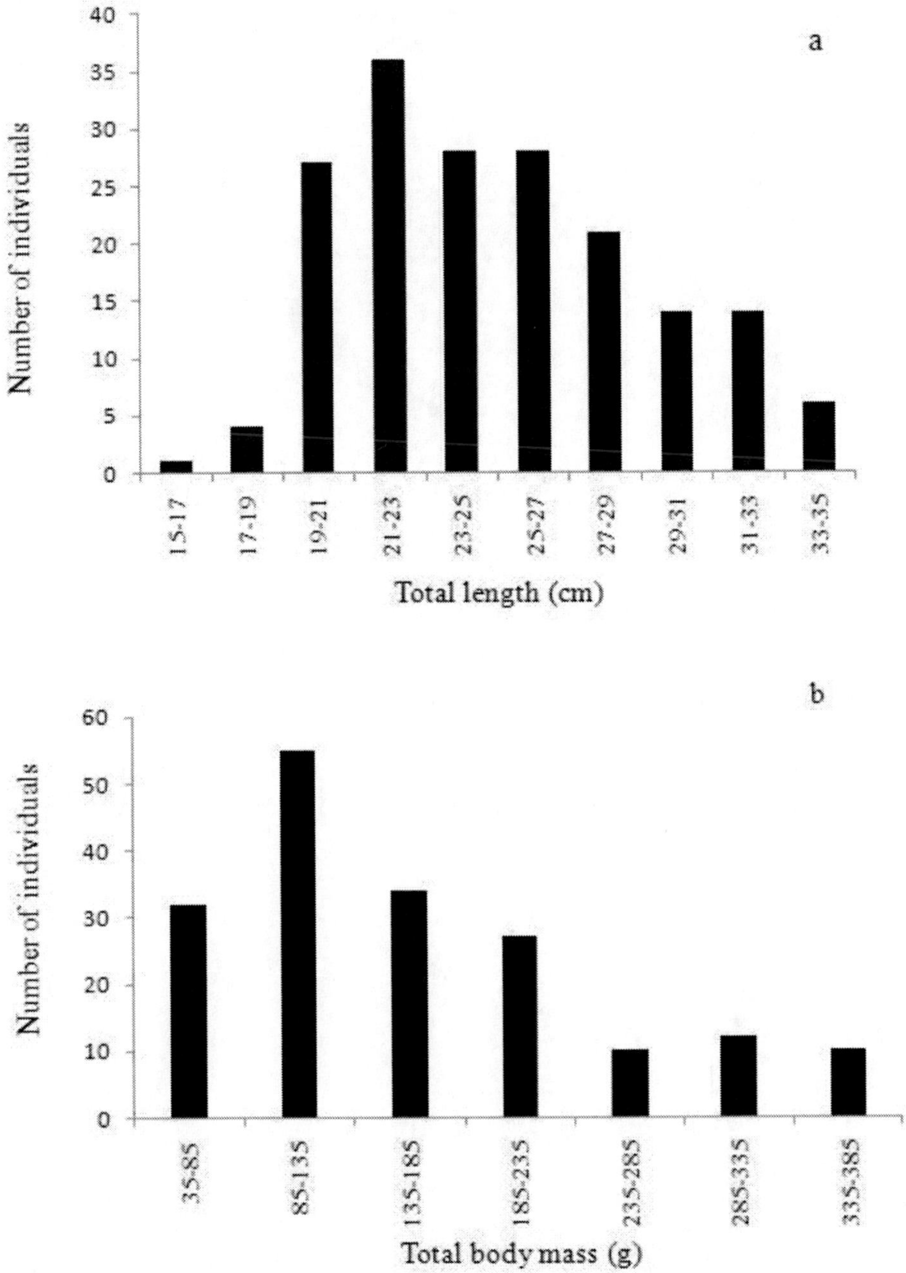

Figure 2. Frequency of occurrence of female *M. curema* by total length classes (a) and body mass (b).

Length-Weight Relationship of Females

The resulting function of L-W relationships for females were $W = 0.0099L^{2,985}$, ($r^2 = 0.97$) (Figure 3), with a positive correlation between total length and body mass, thus presenting a coefficient of 2.985 indicating isometric growth.

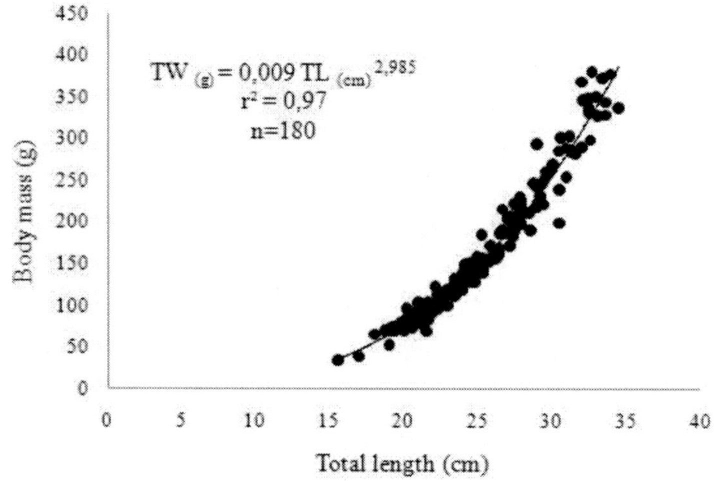

Figure 3. Length-Weight relationship of females of *M. curema*.

Length at First Maturity (L_{50}) of FEMALES

The analysis of cumulative frequency distribution by total length for the adult samples of *M. curema* females revealed that the minimum length at first sexual maturity (L_{50}) was 24. 3 cm of total length (Figure 4).

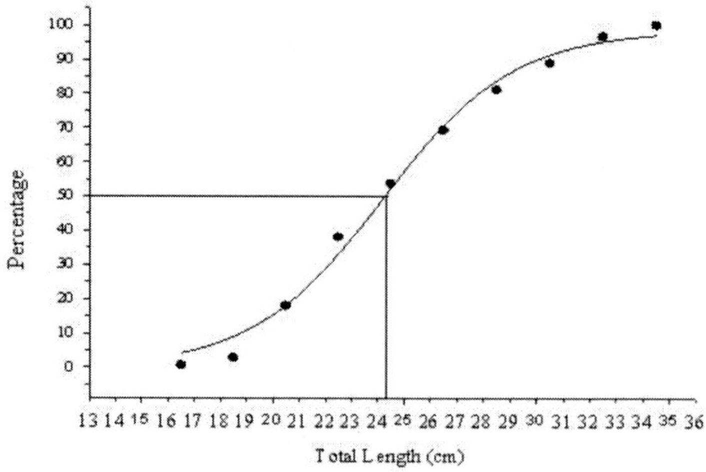

Figure 4. Length at first sexual maturity of female *M. curema*.

Macroscopic Description of Ovaries

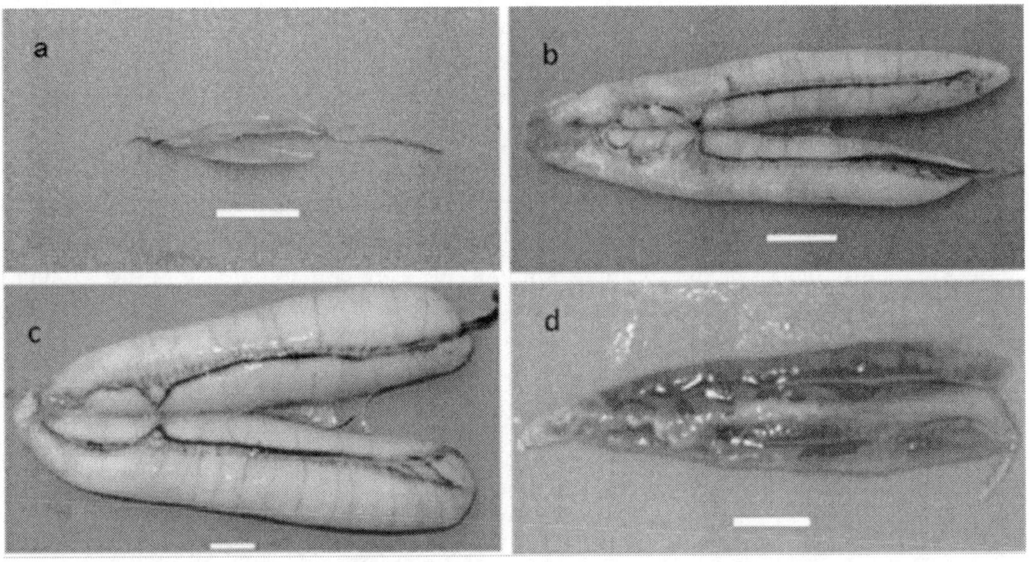

Figure 5. Macroscopic stages of ovarian development of *M. curema* (a) immature, (b) maturing, (c) mature, (d) spent. (Scale: 1 cm).

The ovaries of *M. curema* presented as paired structures, elongated and lobulated with weights ranging from 0. 004 g to 48.5 g (3.0 ± SD 8.34 g). The shape, size, weight, blood vessels and staining varied during the ovarian development (Table 1). In the macroscopic observations of the ovaries the following four developmental stages were observed: immature, maturing, mature and spent (Table 1) (Figure 5).

Histology and Oocyte Development

Five stages of ovarian development were determined microscopically for *M. curema* according to each growth stages of oocytes (Table 1) (Figure 6). The development of oocytes was classified into two phases.

The first phase of development or pre-vitellogenic phase includes the young germ cell (stage I) and perinucleonar (stage II). The second phase of development or the vitellogenic phase, includes lipid vitellogenesis (stage III), vitellogenesis of lipid and protein (stage IV) and post-vitellogenesis (stage V).

Young germ cell or chromatin nucleolus stage **(**Stage I). The smallest cells were found (<10 µm), which appear grouped in nests placed in the ovigerous lamellae, with little cytoplasm, usually with a single central rounded nucleolus, intensely basophilic. These cells were observed in immature ovaries (Figure 6 a).

Early perinucleolus stage or reserve stock (Stage II). The nucleus was in the centre, initially with one or two spherical nucleoli (intensely basophilic) and the cytoplasm was well defined and more basophilic than in the previous phase, measured 10-50 µm.

Late perinucleolus stage (Stage II). The nucleoli become more numerous and were present in the periphery of the nucleus. The cytoplasm is less basophilic showing oocytes of all stages of maturity, measuring 40-70 µm (Figure 6b).

Lipid vitellogenesis (Stage III). The oocytes measured 70-150 µm, showing the central nucleus, cytoplasm is less basophilic than in the previous phase, and with a vacuole representing the lipid deposition. Found mainly in early maturing ovaries (Figure 6b).

Lipid and protein vitellogenesis (Stage IV). The oocytes measured 170-250µm, showing the deposition of protein in the form of platelets from peripheral acidophilic cytoplasm. This stage was present in final maturation of the ovaries (Figure 6c). Post-vitellogenesis (Stage V). The oocyte rapidly increases in size, depending on the accelerated increase in the number of acidophilic yolk granules. The diameter of the oocytes was from 320 to 510µm. This stage was present in mature ovaries (Figure 6d).

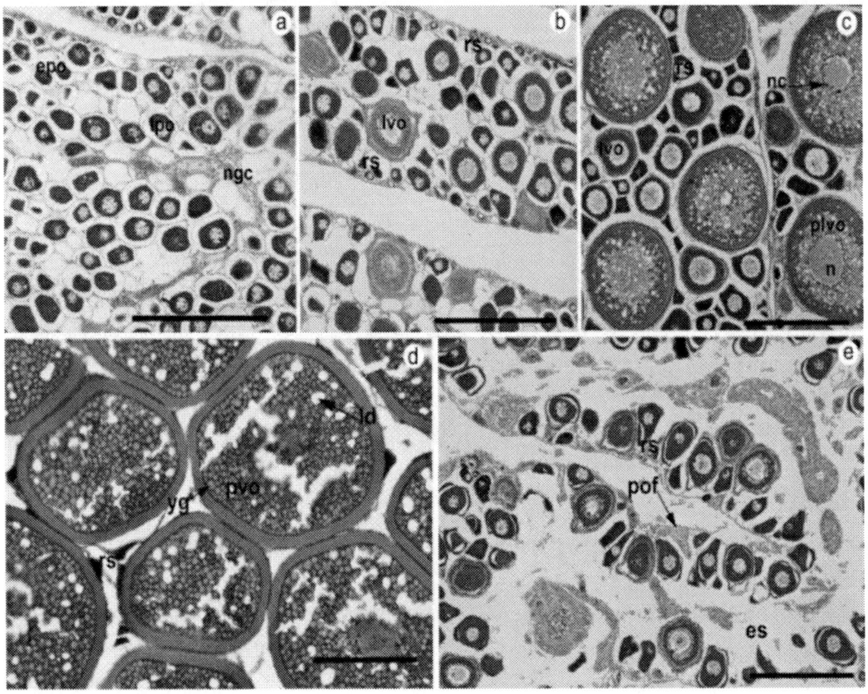

Figure 6. Histological aspects of oocytes developmental stages of *M. curema*: (a) immature ovary containing nest of germ cells (ngc), early perinucleolus oocytes (epo) and late perinucleolus oocytes (lpo); (b) early maturing ovary containing lipid vitelogenic stage oocytes (lvo); (c) maturing ovary containing lipid and protein vitelogenic stage oocytes (lpvo); (d) mature ovary containing post-vitellogenic oocytes (pvo) with yolk granules (yg) and lipid droplets (ld) and (e) spent ovary containing post-ovulatory follicle (pof). rs, reserve stock oocytes; es, empty spaces; n, nucleus; nc, nucleolus. (scale bar = 200µm).

Fecundity and Type of Spawning

Absolute fecundity estimated for mature ovaries was 245,828 oocytes. Relative fecundity was estimated at 814 oocytes per gram body mass of the fish. A high frequency of oocytes with complete vitellogenesis along with perinucleolar oocytes, were observed in histological sections of mature ovaries of *M. curema*. The presence of two batches of oocytes was observed in mature ovaries: the reserve stock of ovogonia and mature oocytes, which indicate the synchronous oocyte development in two groups, characteristic of total spawners.

Table 1. Macroscopic and histological classification and description of the ovarian maturity stages of female *M. curema*

Maturity stage	Macroscopic description	Microscopic description
Immature	Ovaries are small, translucent and oocytes were not visible to the naked eye (Fig. 5a).	Nest of germ cells (oogonia) compactly fill the ovaries and clusters of very small oocytes are beneath the ovigerous lamella (Fig. 6a).
Maturing	The ovaries are larger, lobulated in shape, with a slight reddish tinge and occupying about 1/3 to 2/3 of the coelomic cavity. Blood vessels and small opaque oocytes were visible to the naked eye (Fig. 5b).	Early stage: occurrence of early perinucleolus and late perinucleolus oocytes. Initially the nucleus is with a nucleolus and later appears numerous nucleoli. Oocytes are in previtellogenic stage (Fig. 6b). Late stage: oocytes are mostly with yolk granules and oil vesicles. Reserve stock of germ cells and oocytes with lipid and protein granules are present (protein granules are pink in colour) (Fig. 6c).
Mature	Ovaries are turgid in appearance, orange in colour and occupy 1/2 of the coelomic cavity. Numerous larger opaque oocytes with intense superficial vascularization. (Fig. 5c).	Mature oocytes are predominant, with yolk plates. Reserve stock of germ cells separated in nests are present (Fig. 6 d).
Spent	Ovaries are elongated, flaccid in appearance and reddish-brown in colour (Fig. 5c).	Ovaries show hemorrhaging areas and empty spaces. Reserve stock of germ cells and residual oocytes in the reabsorbing process are present (Fig. 6e).

Frequency of Occurrence of Ovarian Maturity Stages

The frequency of occurrence of ovarian developmental stages showed that females in the immature and maturing stages occurred throughout the year. Mature females were observed from October, 2008 to July 2009, excepting the month of December. Spawned individuals were registered during the months of October, 2008 to July 2009, excepting the months of November and March (Figure 7).

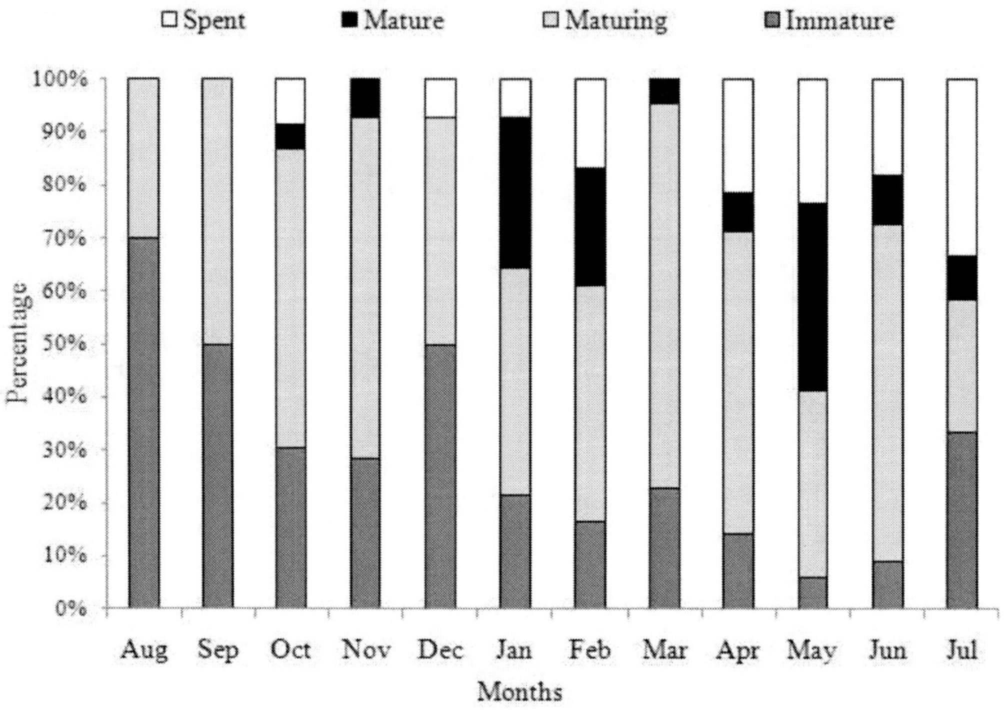

Figure 7. Monthly frequency of ovarian maturity stages in *M. curema* captured during the period August, 2008 to July, 2009.

GSI, **Reproductive Period and Rainfall**

The GSI of females ranged from 0.20 to 4.23 (1.23 ± 1.21). During the study period rainfall varied from a minimum of 0.4 mm in December, 2008 to a maximum of 427.9 mm in August, 2008 (208.2 ± 154.8). Northeastern Brazil experienced 4 months of drought from September to December (19.9 ± 17.7), and 8 months of rain from January to August (302.3 ± 84.6).

The highest values of GSI of females were registered during the rainy season and lowest during the dry season (ρ =0,35; p = 0,243) (Figure 8). The presence of mature individuals during October to July and increased values of GSI, specifically in January, February and May 2009, shows that females have a prolonged period of reproductive activity, with two clear peaks, the first in January-February and the second in May, coinciding with increased rainfall.

DISCUSSION

Females of *M. curema* in smaller classes of total length were abundant in the coastal waters where they were captured. This possibly suggests that this ecosystem offers adequate conditions for the development and survival of the juveniles. Quiñonez-Velázquez and López-Olmos (2011) observed that early juveniles of white mullet have greater growth rates in the coastal area, but with greater risks of mortality. The growth of juveniles is more rapid in the coastal zone than in the lagoon, and the strategy of moving to protected areas is to reach a length that maximizes escape from predation when they return to the coastal areas where the conditions for growth are better and probability of survival is maximized.

Growth is an important factor for establishing optimal fishery management and particularly for understanding the population dynamics of species (Ferrer Montaño and Villasmil, 2008). It is possible to determine the type of growth of a species using the allometric coefficient (*b*) (Jobling, 2002). The L-W relationships of *M. curema* indicate isometric growth within the expected range of 2.5-3.5. The parameters of L-W relationships in fish are affected by factors such as environmental conditions, gonad maturity, sex, stomach fullness, health condition, season, population and differences within species (Froese, 2006).

For the rational management of a fishery stock subject to exploitation, it is of basic importance to know the size at first gonadal maturation (L_{50}), since it provides information for determining the minimum size at capture and mesh dimensions. The L_{50} value of 24.3 cm obtained in this study is in accordance with that recorded by Marín et al. (2003) for the same species in Venezuela. However, L_{50} value of 27.8 cm for females of the same species was registered by Ibáñez-Aguirre and Gallardo-Cabello (2004). The size at first maturity is not fixed and may vary between individuals of the same species, whose populations are subject to different environmental conditions (Wootton, 1990). The size at first maturity can be anticipated with a better food supply, since this leads to an increase in growth rate (Nikolsky, 1963).

The macroscopic anatomy of the ovaries of *M. curema* conforms to the descriptions of Albieri et al. (2009) for the same species. Solomon and Ramnarine (2007) observed for *M. curema* six macroscopic stages of ovarian development: resting, early maturing, late maturing, mature, partially spawned and spawned. The use of such macroscopic scales contribute to the biological knowledge, but the use of histological analysis and measurement of oocytes reduce the risk of incurring errors (West, 1990).

The five stages of oocyte development and the processes of ovarian development of *M. curema* were verified in this study. Oocytes in the pre-vitellogenic stage were present throughout the reproductive cycle in the ovaries, which are regarded as reserve stock for the subsequent breeding season. When there are two groups of oocytes in the ovaries in maturation, development is considered as synchronous (Wallace and Selma, 1981; Murua and Saborido-Rey, 2003). The present study confirms that *M. curema* has synchronous oocyte development in two groups. However, Somolon and Ramnarine (2007) observed that ovaries of individuals before spawning showed primary oocytes, oocytes with yolk formation and vitellogenic oocytes indicating synchronous development in the three groups.

The value of fecundity estimated for *M. curema* in this study approximates with that of other studies (Ibáñez-Aguirre and Gallardo-Cabello, 2004). A study conducted in the Caribbean for *M. curema* indicates that it is a partial spawner and females spawned twice within a reproductive period (Solomon and Ramnarine, 2007). Present study registers total spawning for *M. curema* in the northeastern coastal waters of Brazil, in agreement with the results of Albieri et al. (2009).

The highest values of GSI and greater occurrence of mature females indicated the reproductive period. The GSI is a very effective indicator to determine the degree of development and reproductive period of fish (Isaac-Nahum and Vazzoler, 1987; Chellappa et al., 2010). This study registered a prolonged reproductive period for *M. curema*, with peaks coinciding with increased rainfall. In general, inshore species in lower latitudes exhibit a longer spawning period (Longhurst and Pauly, 1987). Marín et al. (2003) recorded mature individuals of *M. curema* throughout the year in the island of Margarita, Venezuela and successful spawnings coincided with the end of the rainy season. In tropical environments, temperature and photoperiod have low seasonal variations, however seasonal changes of wind and rain cause some seasonality (Lowe-McConnell, 1987). In tropical regions the rainfall plays an important role in determining the reproductive cycles of fish and collective reproduction occurs during the time when environmental conditions are favourable for the survival of juvenile forms and when adequate food is available, besides protection from predators (Chellappa et al., 2010). *M. curema* has a prolonged reproductive period, with two peaks of reproductive activity in January-February and in May, coinciding with increased rainfall.

ACKNOWLEDGMENTS

The authors wish to thank the Federal Post-Graduate Agency of Brazil (CAPES/MEC) (Scholarship granted to M.R. Oliveira) and the National Council for Scientific and Technological Development of Brazil (CNPq) for the Research grants awarded during the study period (E.F.S. Costa and S. Chellappa).

REFERENCES

Albieri, R.J., Araújo, F.G. and Ribeiro, T.P. (2009). Gonadal development and spawning season of white mullet *Mugil curema* (Mugilidae) in a tropical bay. *Journal of Applied Ichthyology, 26*, 105-109.

Albieri, R.J., Araújo, F.G. and Uehara W. (2010). Differences in reproductive strategies between two co-occurring mullets *Mugil curema* Valenciennes, 1836 and *Mugil liza* Valenciennes, 1836 (Mugilidae) in a tropical bay. *Tropical Zoology, 23*, 51-62.

Chellappa, S., Lima, J.T.A.X., Araújo, A. and Chellappa, N.T. (2010). Ovarian development and spawning of Serra Spanish mackerel in coastal waters of Northeastern Brazil. *Brazilian Journal of Biology*, *70* (2), 631-637.

Ferrer Montaño, O.J. and Villasmil, L. (2008). Growth and reproductive aspects of white mullet, *Mugil curema,* in Lake Maracaibo, Venezuela. *Ciencia 16*, (1) 10-19.

Froese, R. (2006). Cube law, condition factor and weight-length relationshipd: history, meta-analysis and recommendations. *Journal of Applied Ichthyology, 22,* 241-253.

Harrison I.J. (2003). Mugilidae. In P.J. Miller (Ed.). *The Freshwater Fishes of Europe*, (8, pp. 1-42). Germany: Wiebelsheim, AULA-Verlag.

Harrison, I.J. (2002). Mugilidae. Mullets. In K.E. Carpenter (Ed.), *The living marine resources of the western central Atlantic: Bony fishes part 1 (Ascipenseridae to Grammatidae)* (2, pp. 1071-1085). FAO Species Identification Guide for Fishery Purposes, American Society of Ichthyologists and Herpetologist Special Publication 5, Rome, 1373 pp.

Hayes, D.B., Brodziak, J.K.T and O'Gorman, J.B. (1995). Efficiency and bias of estimators and sampling designs for determining length-weight relationships of fish. *Canadian Journal of Fisheries and Aquatic Sciences,* 52, 84-92.

IBAMA. (2007). Estatística da pesca 2007. In: G.M. Oliveira; G.C.B. Santos; J.B. Silva Filho; J.A.N. Aragão and S.N. Bezerra. (Eds.), *Brasil: grandes regiões e unidades da federação*, (113p). Brasília, Ministério do Meio Ambiente, Relatório técnico.

Ibáñez-Aguirre, A.L. and Gallardo-Cabello, M. (2004). Reproduction of *Mugil cephalus* and *M. curema* from a coastal lagoon to the northwest of the Gulf of Mexico. *Bulletin of Marine Science,* 75, 37–49.

Isaac-Nahum, V.J. and Vazzoler A.E.A.M. (1987). Biologia reprodutiva de *Micropogonias furnieri* (Desmarest, 1823) (Teleostei, Sciaenidae), 2. Relação gonadossomática, comprimento e peso dos ovários como indicadores do período de desova. *Boletim do Instituto Oceanográfico,* 35, 123-134.

Jobling, M. (2002). Environmental factors and rates of development and growth. In P.J.B. HART and J.D. REYNOLDS (Eds.), *Handbook of fish biology and fisheries* (1, pp. 97-122). Oxford: Blackwell Publishing.

Katselis, G., Koutsikopoulos C., Rogdakis, Y., Lachanas, T., Dimitriou, E. and Vidalis, K. (2005). A model to estimate the annual production of roes (avgotaracho) of flathead mullet (*Mugil cephalus*) based on the spawning migration of species. *Fishery Research,* 75:138-148.

Lessa, R.P. (2006). Recursos pesqueiros da região nordeste. In Ministério do Meio Ambiente (MMA). *Avaliação do potencial sustentável de recursos vivos na Zona Econômica Exclusive* (pp. 153-180). Brasília: MMA. (Programa REVIZEE-SCORE-NE, Relatório Executivo).

Longhurst, A.R. and Pauly, D. (1987). *Ecology of Tropical Oceans*. London: Academic Press.

Lowe-Mcconnell, R.H. (1987). *Ecological studies in tropical fish communities.* Cambridge: Cambridge Tropical Biology Series.

Marín, B. J., Quintero, A., Bussière, D. and *Dodson, J. J. (2003).* Reproduction and recruitment of white mullet (*Mugil curema*) to a tropical lagoon (Margarita Island, Venezuela) as revealed by otolith microstructure. *Fishery Bulletin,* 101, 809–821.

Menezes, N.A. (1983). Guia prático para conhecimento e identificação de tainhas e paratis (Pisces, Mugilidae) do litoral brasileiro. *Revista Brasileira Zoologia,* 2, 1-12.

Moreno, T., Castro J.J. and Socorro J. (2005). Reproductive biology of the sand smelt *Atherina presbyter* Cuvier, 1829 (Pisces:Atherinidae) in the central-east Atlantic. *Fisheries Research,* 72, 121–131.

Murua, H. and Saborido-Rey, F. (2003). Female Reproductive Strategies of Marine Fish Species of the North Atlantic. *Journal* of *Northwest Atlantic Fishery Science,* 33, 23-31.

Murua, H., Kraus, G., Saborido-Rey, F., Witthames, P.R, Thorsen A. and Junquera, S. (2003). Procedures to Estimate Fecundity of Marine Fish Species in Relation to their Reproductive Strategy. *Journal of Northwest Atlantic Fishery Science,* 33, 33-54.

Nelson, J.S. (2006). *Fishes of the World* (4). New York, NY: John Wiley and Sons, Inc.

Nikolsky, G.V. (1963). *The ecology of fishes*. London: Academic Press.

Pina, J.V. and Chaves, P.T. (2005). A pesca de tainha e parati na Baía de Guaratuba, Paraná, Brasil. *Acta Biológica Paranaense,* 34: 103-113.

Quiñonez-Velázquez, C. and López-Olmos, J.R. (2011). Juvenile growth of white mullet *Mugil curema* (Teleostei: Mugilidae) in a coastal lagoon southwest of the Gulf of California. *Latin American Journal of Aquatic Research*, 39(1), 25-32.

Santos, M.N., Gaspar, M.B., Vasconcelos, P. and Monteiro, C.C. (2002). Weight-length relationships for 50 selected fish species of the Algarve coast (southern Portugal). *Fisheries Research,* 59, 289-295.

Solomon, F.N. and Ramnarine, I.W. (2007). Reproductive biology of white mullet, *Mugil curema* (Valenciennes) in the Southern Caribbean. *Fisheries Research,* 88, 133–138.

Vazzoler, A.E.A.M. (1996). *Biologia da Reprodução de Peixes Teleósteos: Teoria e Prática.* Maringá: EDUEM.

Wallace, R.A. and Selman, K. (1981). Cellular and dynamic aspects of oocyte growth in teleosts. *American Zoology*, 21, 325-343.

West, G. (1990). Methods of assessing ovarian development in fishes: a Review. *Australian Journal of Marine and Freshwater Research,* 41, 199-222.

Wootton, R.J. 1990. *Ecology of Teleost Fishes*. Londres: Chapman and Hall.

Wootton, R.J., Evans G.W. and Mills, L.A. (1978). Annual cycle in female three-spined sticklebacks (*Gasterosteus aculeatus* L.) from an upland and lowland population. *Journal of Fish Biology,* 12, 331-343.

Yoshida, H.O. (1964). Skipjack tuna spawning in the Marquesas Islands and Tuamotu Archipelago. *Fishery Bulletin,* 65, 479-488.

In: Biology of Semiarid Tropical Fish
Editors: J. R. Gomes and S. Chellappa

ISBN: 978-1-63117-067-6

Chapter 3

HISTOLOGICAL ASPECTS OF GONAD DEVELOPMENT IN THE DIANDRIC PROTOGYNOUS SEQUENTIAL HERMAPHRODITE FISH *SYNBRANCHUS MARMORATUS* (OSTEICHTHYES: SYNBRANCHIDAE)

***Nirlei Hirachy Costa Barros*[1], *Arrilton Araújo de Souza*[1] *and Sathyabama Chellappa*[2*]**

[1]Postgraduate Programe in Psychobiology, Centro de Biociências, Universidade Federal do Rio Grande do Norte, Av. Salgado Filho, Lagoa Nova, Natal, Rio Grande do Norte, Brazil

[2]Department of Oceanography and Limnology, Centro de Biociências, Universidade Federal do Rio Grande do Norte, Natal, Rio Grande do Norte, Brazil

ABSTRACT

The reproductive system of the marbled swamp eel, *Synbranchus marmoratus* has striking differences from the other fish, due to its sequential hermaphroditism. This study investigated the histological aspects of gonad development of this species captured from a tropical reservoir in Northeastern Brazil. The sampled population consisted of primary males, females, transitional individuals and secondary males. Primary males develop directly as males while secondary males arise from the sex reversal of females. The fish captured were identified as to sex and developmental stages using macroscopic and histological techniques. Four different sexual types were identified macroscopically, with gonads showing variations in shape, coloration and presence or absence of blood vessels. Of the 179 individuals examined, 12 were primary males, 126 were females, 14 were transitional individuals and 27 were secondary males. Histological analyses revealed the gonad structures in the different sex types and changes which occurred during the sex reversal of females to secondary males. *S. marmoratus* is a diandric, protogynic hermaphroditic fish. The females were sexually active with 43.5 cm of total body length and the secondary males with 58.5cm of total length. During the process of sex reversal in *S. marmoratus*, the female suffers several specific internal changes, where the ovarian

* E-mail of corresponding author: chellappa.sathyabama63@gmail.com.

tissue is totally replaced by male tissue and this process only occurs in adults. *S. marmoratus* presents total spawning with a mean fecundity of 497.0 (±153.3).

Keywords: *Synbranchus marmoratus*, histology, primary males, females, transitional individuals, secondary males

INTRODUCTION

The teleost fish show a great variability of reproductive mechanisms, ranging from gonochorists (individuals with separate sexes) to hermaphrodites (individuals with both sexes). The latter, which is the expression of both male and female reproductive function in a single individual, generates curiosity and controversy (Chan and Yeung, 1983; Reinboth and Brusle-Sicard, 1997; Devlin and Nagahama, 2002). Attention has been attracted with reports of manifestations of the enormous plasticity in sexual pattern expressed by teleosts in the form of bi-directional sex change in adults and the influence of social factors on sexual development in juveniles (Munday et al. 2006). According to sex allocation theory, individuals in hermaphroditic fishes should change sex whenever net future reproductive success would be higher for the opposite than for the existing sex (Shapiro, 1987).

The normal occurrence of spontaneous sex inversion among teleosts is a recent issue within current concepts on vertebrate reproductive physiology. Investigations on this subject are mostly descriptive and deal with morphological and behavioural aspects. The hermaphroditic fish can be divided into two types, being simultaneous and sequential. In simultaneous hermaphroditic fish the individual presents both sexes functional at the same time, however, this form is not common among teleosts (Reinboth, 1983). The sequential hermaphrodites are widespread among teleosts, and this modality of sex inversion may be a spontaneous event in the species and or controlled by genetic, physiological and behavioral factors (Sadovy and Shapiro, 1987).

The sequential hermaphroditism may be either protandry or protogyny. Protandrous species change sex from adult male to female, whereas protogynous species change sex from adult female to male. Within the protogynic hermaphroditism, the species may have two standards related to males, the monoandria (individuals which are born and function as a males throughout their lifetime) and diandria (two types of males: primary males, which are born and function as males throughout their lifetime; and secondary males, which are born as females and later become males due to sex reversal) (Reinboth, 1983).

Studies related to the sex reversal process treated fish species of the family Synbranchidae as the main target (Lo Nostro and Guerrero, 1996; Lo Nostro, 2000). *Synbranchus marmoratus* Bloch 1975 is distributed from Argentina (south of Buenos Aires province) to Guatemala and Mexico, and throughout Brazil (Britski et al., 1999). *S. marmoratus* captured in Argentina demonstrated that this species is a diandric, protogynic sequential hermaphrodite, presenting four sex types in its population: primary males, females, transitional individuals and secondary males (Lo Nostro and Guerrero, 1996). It is popularly known as the marbled swamp eel, has a serpentine body without scales and is commonly found in the Brazilian semiarid region. This species has developed mechanisms to adapt to the dry conditions of this region and its response to long periods of drought is remarkable. They bury themselves in the river or reservoir beds by constructing mucus cocoons surrounding them, and their metabolic processes are greatly reduced and they enter a state of aestivation, a form of extreme hibernation (Montenegro et al., 2011; Montenegro et al., 2012).

Fish living under semi arid climatic conditions of Northeastern Brazil are faced with several problems, and the most obvious of which is the lack of water. During the dry seasons, when the rivers and reservoirs dry up or have limited water sources, *S. marmoratus* is subjected to increasing environmental stress. Although *S. marmoratus* has no direct commercial importance, it is considered as a good source of protein for the fisherman community in this region, especially because of its abundance in reservoirs (Barros, 2012).

In this study of the marbled swamp eel, *S. marmoratus* the following questions were addressed:

1. What are the histological changes that occur in the gonads during the sequential hermaphroditic reproductive cycle?
2. What is the size at first sexual maturity, fecundity and type of spawning?

The answers of these questions describe the histological aspects of gonads of primary males, females, transitional individuals, secondary males, the size at first sexual maturity, fecundity and type the of spawning of the *S. marmoratus* population sampled from Northeastern Brazil. This information could be useful for the understanding of the plasticity in sexual development of *S. marmoratus.*

MATERIALS AND METHODS

Study Site and Fish Sampling

Individuals of *S. marmoratus* were captured on a monthly basis from July, 2011 to July, 2012 from the Marechal Dutra Reservoir, State of Rio Grande do Norte, located in Northeastern Brazil (6° 26' 24''S and 36° 38' 00''W). This reservoir has an accumulation capacity of 40×10^6 m^3 with maximum depth of 25m and mean of 8.5m, and is a source of potable water, besides being important for fisheries and agriculture. This reservoir was constructed on the River Acauã in the hydrographic basin of River Piranhas-Assu, which is the main hydrographic basin of Rio Grande do Norte, covering an area of 44,000 km^2 (Chellappa et al., 2009).

Fish samples were captured with the help of local fishermen, who used artisanal fish traps locally known as "covos" which were placed on the substrate close to the marginal vegetation. The fish traps were allowed to remain there stationary for 24 hours. The captured fish were numbered, measured (total length Lt $\pm$ 1cm, from the oral to caudal end) and weighed (total body mass Wt $\pm$ 1 g and gonad weights of Wg $\pm$ 1 g). Some fish were used for morphometric measurements to confirm the taxonomical identification of the study species (Britski et al., 1999).

Macroscopic and Histological Examinations of Gonads

The location and general aspects of the gonads were observed and stages of reproductive maturity were determined using a macroscopic staging system. The macroscopic characteristics used for sex differentiation of the individuals were based on the external morphology of the gonads, such as, the size, form, coloration, presence of blood vascularization, presence of visible oocytes and the amount of coelomic space occupied by the gonads. The sexes were separated and the gonads selected for histological procedures (n= 40) were fixed in Bouin solution for 12 to 24 h, washed for 24hs in water and preserved in 70% Ethyl alcohol. The ovaries and testicles were later embedded in paraffin wax, sectioned at 3-5 μm thickness, and stained with Harris Hematoxylin and Eosin (H&E) (Smith et al., 1996).

In order to avoid possible variation in the developmental stage of oocytes due to their position in the ovaries, histological examinations were carried out on sections from the anterior (cephalic), middle (central), and posterior (caudal) regions of 40 ovaries in different developmental stages. These data were later compared in order to determine whether samples taken from mid-section of the ovary were representative of oocyte development (Yoshida, 1964). Histological verification indicated that gonads were at the same stage of maturation throughout their length. Subsequently, only a central transverse section was prepared for each gonad (n= 40). Stages of gonad maturation were determined using a light microscope (Taimin, model TM 800), coupled with a video camera (Kodo Digital). Existing terminology was used for staging of oogenesis (Wallace and Selman, 1981).

Sex Ratio and Length at First Gonadal Maturity (L_{50})

After a longitudinal incision in the abdomen of each fish the gonads were removed, and the macroscopic aspects and maturation stages of the gonads were observed besides determining the sex of each fish.

L_{50} was defined as the smallest predicted length interval at which 50% of the individuals had maturing gonads. This was estimated from the relative frequency distribution of adult fish, using their total length classes (Ls mean ± SD) (Moreno et al., 1977).

Fecundity and Type of Spawning

To determine the total fecundity, ovaries were removed from eight mature females, weighed and preserved in Gilson solution for 24 hours for complete dissociation of oocytes, which were then washed and preserved in 70% Ethyl alcohol (Gomiero et al., 2009). Triplicates of 10% sample were removed for counting of mature oocytes and the values were extrapolated to 100%. The diameters of oocytes were measured with an ocular micrometer (Wild M7, objective 31x and ocular 10x).

Spawning pattern was assessed based on the oocyte diameter distributions. The oocytes of mature ovaries were grouped into classes of different diameter sizes, based on their frequency of occurrence, to determine the type of spawning (West, 1990).

Statistical Analyses

Distributions of raw data were checked for normality and transformed when necessary. The t-test was used to determine the statistically significant differences between the lengths and weights. The chi-square (χ^2) test was applied to check the differences between the sex ratio established. Student t test was used to verify the difference between the total length and body mass of fish, at a significance level of 5%.

RESULTS

A total of 179 individuals of *S. marmoratus* were captured and it was observed that the sampled population constituted of four different sex types: primary males (born and function as males; n=12), functional females (n=126), transitional individuals (individuals with both male and female gonadal tissues in the process of changing sex; n=14) and secondary males (females transformed as males; n=27).

Excluding transitional individuals, there was a predominance of females during the entire sampling period (χ^2=57.51; p = 0.0001). The primary males of *S. marmoratus* showed amplitude of total length varying from 22 to 32 (26.3±3.95) cm, the females from 22 to 56 (43.9± 6.85) cm, transitional individuals from 37 to 56 (49.3±6.7) cm and secondary males from 45 to 68 (57±4.9) cm. The secondary males were the largest and the primary males were the smallest. The highest frequency of females and transitional individuals were in the class range of 42 to 52 cm. The amplitude of total weight of the primary males of *S. marmoratus*, varied from 6 to 50 g (21.6±16.94), females from 9 to 210g (101.4 ±44.5), transitional individuals from 55 to 222g (148.3±48.3) and secondary males from 32 a 327g (203±70.3). The secondary males were the heaviest and the primary males had the lowest weights. The highest frequency of females and transitional individuals were in the class range of 66 to126g.

Morphological Characteristics that Differentiate the Sex of *S. marmoratus*

Primary Males

These had paired bilobular testes, which were located along the mid-dorsal sides of the body cavity and occupied approximately two-thirds of the total body length. They appeared like two cords on either side, which were joined medially by connective tissue. The posterior end was at the urogenital papilla which opened behind the anus. They were translucent to whitish in color and were elongated and cylindrical in mature males or filiform in immature males. A small depression in the form of a line was observed in the center of each testis following its whole length, which was the efferent duct (Figure 1a). Testis structure conforms to the unrestricted lobular type. All secondary males examined were immature.

Females

Morphologically, the female had an unpaired ovary which was long, cylindrical with rounded edges. The maturation of the ovary was followed by various changes which were observed by the naked eye: the coloration, size, form and volume. These changes were observed in all three stages of development, young, maturing and mature females. Females in the young stage, had one translucent unpaired ovary, in the form of a ribbon which occupied a space lesser than 1/3 of the coelomic cavity. The size and volume were relatively small when compared to the other stages. In the maturing stage, the ovary was well irrigated by blood vessels and was reddish to deep pink in color, also showing increase in size and volume. The mature ovary was considerably big occupying almost all the available space of the coelomatic cavity, was yellowish orange in color and big oocytes were present in the whole ovary (Figure 1b).

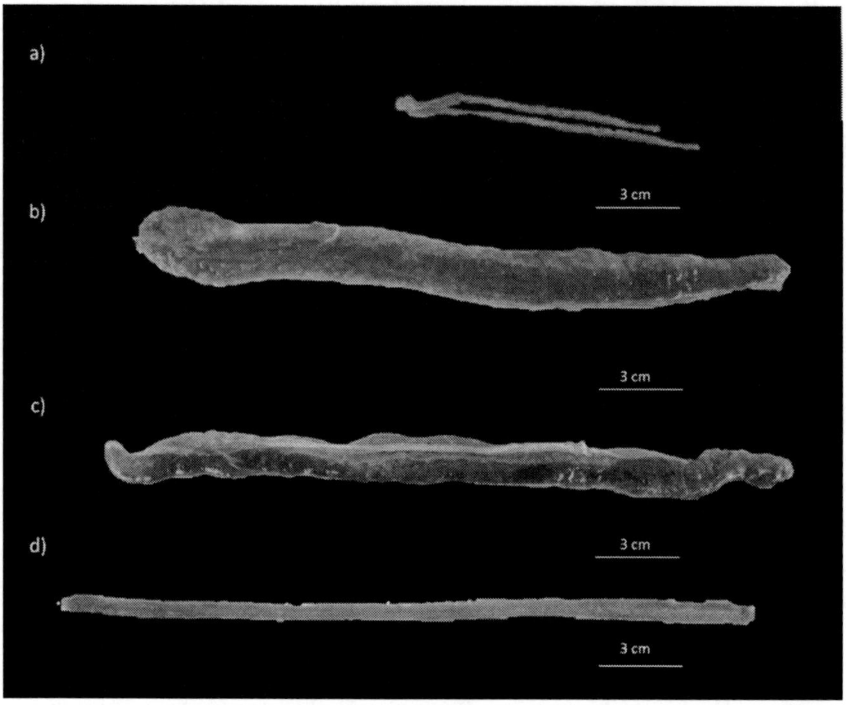

Figure 1. Macroscopic visualization of the gonads of *S. marmoratus*. The gonads of (a) primary male, (b) female, (c) transitional individuals and (d) secondary male (Scale bar = 3 cm).

Transitional Individuals

The transitional individuals originate from the females when they enter the sex reversal phase and hence, the gonad appears as a thick longitudinal cord like structure, located in the cephalo-caudal direction. The ovary is cylindrical, occupying a large space of the coelomic cavity, with visible oocytes in the central part. In the limiting region of the ovarian cavity, there were dense groups of cells, white to pale yellow in color, indicating cellular disorganization and the appearance of testicular tissue. The later substitutes the female germinative tissue, which is the process of sex reversal (Figure 1c).

In the initial stage of transition, the ovary was dark pink in the central region and yellowish in the limiting region of the ovarian cavity, with intense blood vessels, visible oocytes, with morphological resemblance to the ovary of a functional female. Later the form of the gonad remained unchanged, but with lesser oocytes than in the earlier stage. The cellular disorganization was more evident and the testicular tissues invaded and substituted the ovarian tissues. The gonad became light yellowish in color, and was characterized by the presence of a large number of visible black dots along the entire length of the gonad. In the final stage, the gonad became thin and flat, with an even color, and without oocytes, thus taking the form of a testis.

Secondary Males

Testis of secondary males develops within the single ovary of a female during sex reversal. The anatomical shape of the male gonad is very similar to that of the female except for the spermatic ducts. In secondary male the testes is unpaired, represented by a long, thick and cylindrical cord like structure. At this stage, the sex reversal is completed and the individual becomes a secondary male. The testis contains only the male cells and thus the transformation is complete. The color was almost translucent and the location of the testes was the entire length of the coelomic cavity up to the urogenital papilla. Blood vessels were not evident (Figure 1d).

Histological Changes that occur in the Gonads during the Reproductive Cycle

Primary Males

The testes were surrounded separately by a tunica albuginea, a dense connective tissue and joined at the posterior end by the union of the efferent ducts, forming a single spermatic duct extending up to the urogenital papilla. The testes of primary males showed no sign of an ovarian lumen, as observed in the testis of the secondary males. The primary males captured, were either in an early maturing stage or in a stage of maturing. In the early maturing stage testis, there was a predominance of spermatogonia, the largest cells of the spermatogenic lineage, and the wall of the seminiferous tubules were formed mainly by primary cysts or spermatogonia. In the maturing stage testis, the presence of cells of the spermatogenic lineage, with spermatogonia and spermatocytes were evident. Several secondary spermatogonia were observed within the cysts, which occurred in conjunction with cysts of primary and secondary spermatocytes and cysts with spermatids (Figure 2).

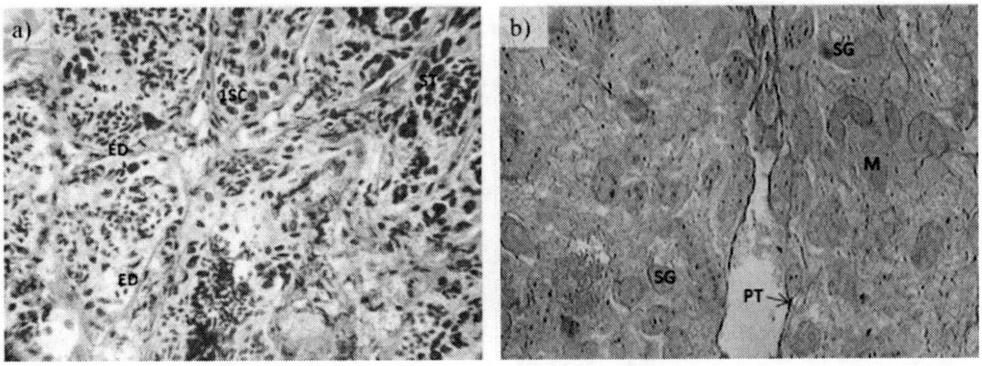

Figure 2. (a) and (b) A cross section the gonad of the primary male of *S. marmoratus*: (ED) efferent ducts, (1SC) sperm in development phase within the spermatocytes, (SG) primary spermatocytes and (M) spermatogonium in metaphase (Scale bar = 100μm).

Functional Females

The ovary of mature female was long and cylindrical, without the medular zone and covered by túnica albuginea, constituted of conjunctive tissue and blood vessels. In the ovigerous lamellae either germinative cells or oocytes in different phases of development (primary, secondary or mature) were present. The ovarian stroma was partly constituted of a prominent extravascular space in the periphery of oocytes in development.

The vitellogenic oocytes had large quantity of yolk globules, the nucleus was smaller than in the earlier stages and the nucleolus was randomly distributed with a narrow zona pelucida. In the ovary of mature functional females the atresic vitellogenic oocytes were observed with a characteristic irregular outline, and were not attached to zona pelucida. The nucleus was desorganized and was migrating towards the extremity (Figure 3).

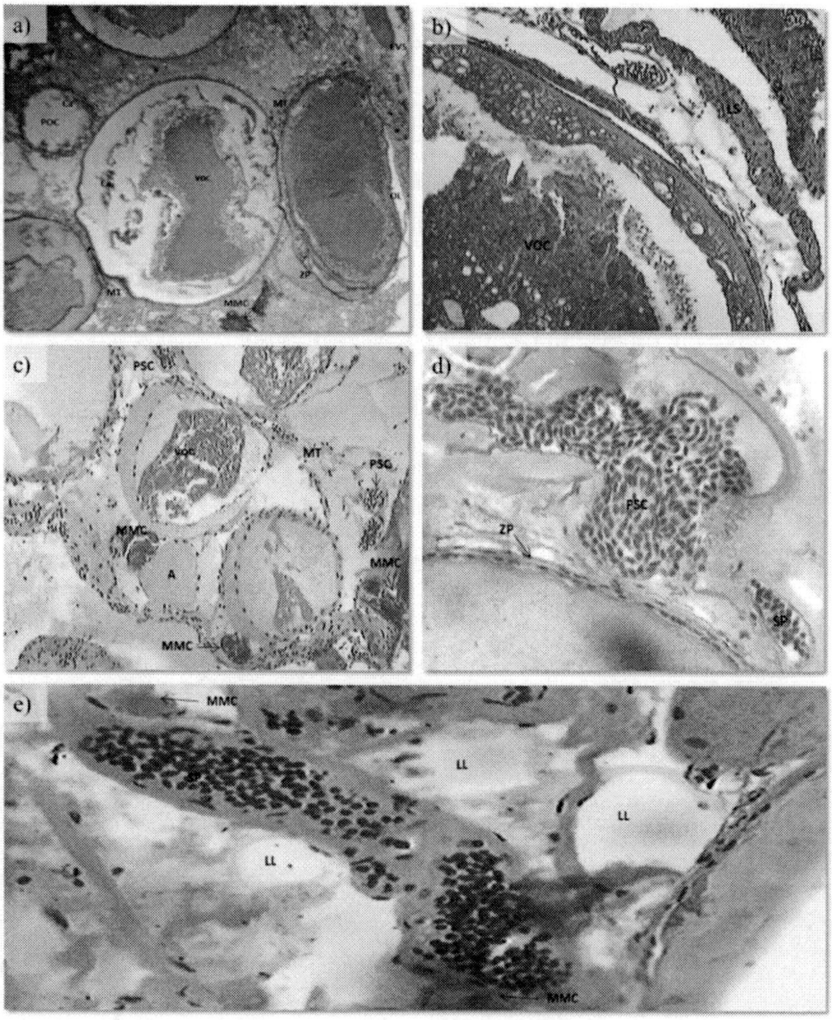

Figure 3. A cross section the gonad of the female *S. marmoratus*. a): (n) nucleus, (ZP zona pellucida, (Ca) cortical alveoli, (POC) primary growth oocytes, (VOC) vitellogenic oocytes, (EVS) extravascular space (scale = 100μm); b) The oocyte vitellogenic (scale = 200μm); c) ovary showing oocytes at different stages (VOC) vitellogenic oocyte, (ZP) zona pellucida, (AVO) atretic oocytes, (ii) chromatin nucleolus (scale = 100μm), d) The atretic oocyte (scale = 200μm).

Transitional Individuals

This phase was characterized by the presence of primary oocytes with cortical alveoli and several vitellogenic oocytes. There were small number of oocytes in the absorption process. The detachment of the zona pellucida was observed in the atretic oocytes, forming an extravascular space. The gonad showed a large proliferation of Leydig cells (also known as interstitial cells of Leydig), and migration of myoid cells. There was invasion of male tissue in the periphery of the ovigerous lamellae between the female germ cells, which were still present in the periphery. In this initial phase melano-macrophagic centres began to appear. These centers were pigmented cell aggregates or black spots, which were displayed next to the germ cells. These centers are mainly responsible for the removal or phagocytosis of cellular debris (Figure 4).

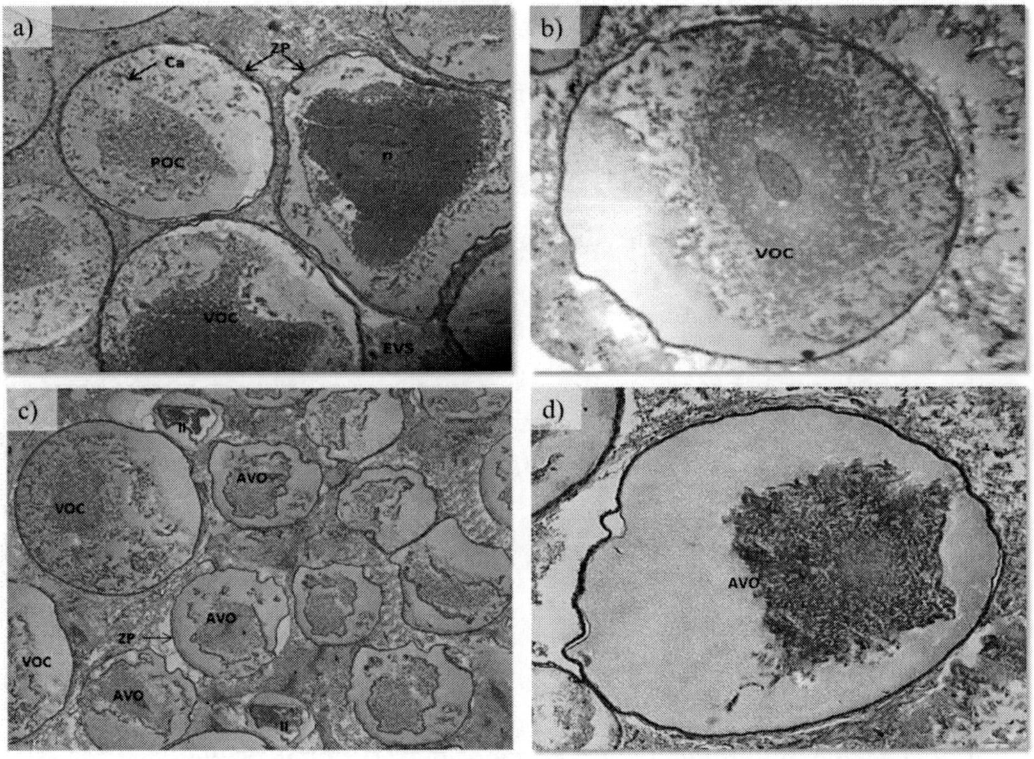

Figure 4. A cross section the gonad of the (a) *S. marmoratus* during the initial phase of transition: (VOC) vitellogenic oocytes, (POC) primary growth oocytes (Ca) cortical alveoli, (MT) male tissue, (OL) lamellae between the oocytes, (EVS) extravascular space, (ZP) zona pellucida, (MMC) melanomacrophage centers (scale = 100μm); (b) and (c) A cross section the gonad during the intermediate stage of intersex: (VOC) vitellogenic oocytes, (LS) ovarian lumen (POC) oocytes in primary growth, (LS) ovarian lumen, (MMC) melanomacrophage centers, (A) atretic bodies, (SP) sperm, (MT) male tissue, (PSC) spermatocytes (scale = 200μm); (d) and (e) A cross section the gonad the gonad during the final phase of transition: (PSC) spermatocytes, (ZP) zona pellucida, (LL) lobe of the lumen devoid of sperm, (MMC) melanomacrophage centers, (SP) sperm (scale = 200μm).

Later, this phase was characterized by having few oocytes and practically all of them in the process of vitellogenic atresia, being gradually absorbed. The male tissue occupied almost 75% of the gonad. The increased presence of melano-macrophagic centre bodies and process of atresia, were more numerous than in the previous phase. The main feature of this stage was the presence of cysts with male cells in various stages of development, viewed as clusters, with spermtocytes (Figure 4).

In the final phase, the gonad was without vitellogenic oocytes and some atretic oocytes were seen in all its extension. All the space formerly occupied by oocytes was filled by male tissue. The presence of many spermatocysts and spermatocytes, as well as the presence of lobes, devoid of sperm were characteristic of this phase. The presence of melano-macrophagic centres was significant, to the total removal of cellular debris of the former ovary (Figure 4).

Secondary Males

In the testes of secondary males the spermatozoa were distributed within the lobules, as several nests throughout the gonad. The lobules were within the lamellae of the former ovary. It was possible to view melano-macrophagic centres in the gonad, as a sign of complete removal of fragments of ovarian tissue. The efferent ducts were visible, organized as channels linked to the lobes. Residual sperm from the previous breeding season was observed. Spermatogonia and spermatozoa were present within the lobules (Figure 5).

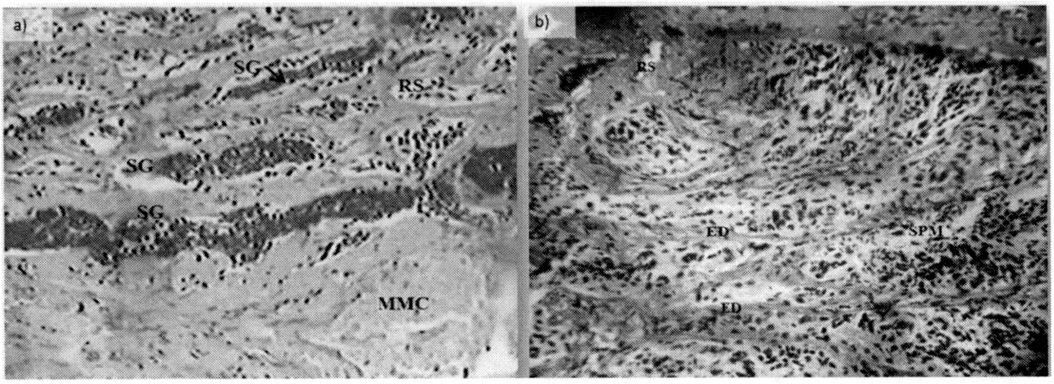

Figure 5. A cross section the testis of secondary male of *S. marmoratus* (a) and (b): (RS) residual sperm, (ED) efferent ducts, (SG) spermatogonium, (SPM) sperm, (MMC) melano-macrophage centers (scale = 100µm).

Size at First Sexual Maturation (L_{50}), Fecundity and Type of Spawning

The size at which 50% of the population of females exhibited the process of gonadal maturation was 39.5 cm of total length. The L_{50} of secondary males was 58.5 cm of total length (Figure 6). All secondary males captured were in the immature to maturing stage.

The fecundity of smaller females (of 42.5 cm length) varied from 311 to 313 mature oocytes, whereas, that of larger females (of 51 cm length) varied from 621 to 627. The mean fecundity was 497±153.3. Histological sections of ovary of *S. marmoratus*, showed the presence of mature oocytes and reserve ovogonia which characterize total spawning.

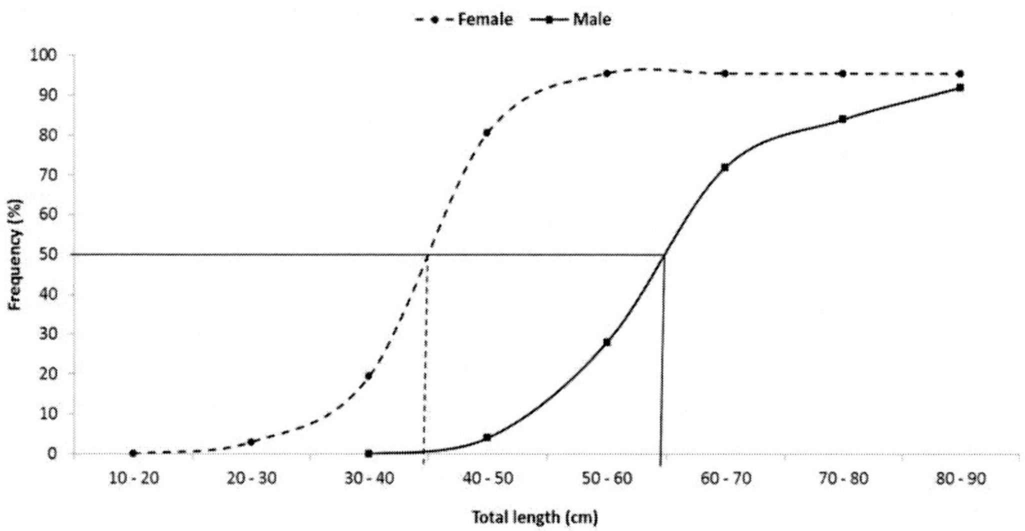

Figure 6. Relative frequency of first gonadal maturation (L_{50}) of *Synbranchus marmoratus*. The inserted lines represent the estimated size range where 50% of the fish were mature.

Discussion

In protogynic hermaphroditism a proportion of individuals can directly differentiate into males, which are known as primary males. Males resulting from the sex reversal of females are called secondary males. A protogynous species that has primary and secondary males are called diandric, and a protogynous species that has only secondary males are called monandric. Besides the females and two types of males, there are transitional individuals, which are characterized by gonads with both feminine and masculine tissues (Reinboth, 1983). In this study, *S. marmoratus* is characterized as diandric protogynic hermaphrodite, because of the presence of primary males in the sampled population. Primary and secondary males may or may not occur in populations of the same species depending on the region (Liem, 1963). The occurrence of two types of males was also encountered in *Monopterus albus* (Liem, 1968).

The macroscopic characteristics of the gonads of *S. marmoratus* did not permit the sex identification, except in primary males which had paired gonads, the right and left lobes. Histological analyses were useful to clearly establish the sex of individuals. The general pattern of the gonads of primary males and females of *S. marmoratus* were similar to that of other gonochorsistic species (West, 1990). For transitional individuals and secondary males, descriptions corroborated with that of Carruth (2000) and Lo Nostro (2000).

Females dominated in number among the individuals captured during this study, possibly because they live in groups with males (Thresher 1984; Moyer 1990). In a group where functional males die or are removed, the larger females undergo sex reversal process and become as functional males. This type of social organization was identified for protogynic hermaphrodites, specifically among the reef fishes, such as, *Paragobiodon* (Thresher, 1984) and *Pomacanthus* (Allen et al., 1998). This behavior is advantages for the *S. marmoratus* population of the semiarid region since spawning occurs during the short spell of the rainy season (Barros, 2012). The primary males do not have a significant role, possibly due to their substitution by the secondary males. This is possibly an evolutionary advantage to have bigger males where there are more females in the population.

Studies on ovogenesis and gonadal classification are very important for the understanding of fish life cycle (Brandão et al., 2003). The gonad maturation process initiates with ovogonia followed by perinuclear oocytes. The endogenic vitollogenesis initiates with the appearance of yolk constituted by glycoprotein or polysaccharides (Hibiya, 1982). The multiplication of vitellogenic oocytes begin during the maturation of the female, and this was observed in the ovary of *S. marmoratus*.

In the female mature gonad of *S. marmoratus* the presence atresic oocytes were visible, characterized by a deregulated zona pellucida, forming a very visible extravascular space. In fish, the components of the zona pellucida are synthesized in the ovary and also in the liver under the influence of estrogen (Murata et al., 1997).

During the process of sex reversal in *S. marmoratus*, the female undergoes several specific internal changes, where the ovarian tissue is totally replaced by male tissue and this process only occurs in adults. In some Labridae fish the process of sex inversion occurs in individuals before sexual maturity (Carruth, 2000). The gonads of transitional individuals is very characteristic, because it shows a relative standard disorganization of the gonad architecture, mass degeneration of female germ cells, the presence of melano-macrophagic centers, invasion of interstitial tissue and male germ cells. In the later phases of transition, cysts containing spermatogonia appear at the edges of the ovigerous lamellae, which give rise to spermatozoids. This histological pattern of was also found in *Monopterus albus* (Liem, 1963). The presence of structures found in the transitional phase of *S. marmoratus* related to the removal, absorption or cleaning of tissue debris, the melano-macrophages were also observed by others (Sadovy and Shapiro, 1987; Grier and Taylor, 1998). In transitional individuals the sexual inversion is marked by the gradual decrease in female activity, degeneration of germ cells, such as unspawned oocytes undergoing resorption. After this process there is an increase in spermatogenic activity.

The secondary males of *S. marmoratus* are the end products of sexual inversion, and present an unrestricted lobular testis, comprising numerous seminiferous tubules which are separated by a thin layer of connective tissue. In the interior of testis are found various cysts with spermatogenic cells in developing stages (Pudney, 1995). Spermatogenesis is synchronic in the early stage of maturation, with cysts containing primary spermatogonia in different sizes. The testis has two separate compartments, the tubular and the interstitial. The tubular part contains the germ and somatic cells that will later form the spermatozoids (Billard et al., 1982).

Protogynous species are more susceptible to overfishing because they need to reach a maximum size for sex reversal. If the fishing pressure does not permit individuals to reach the size required for sex reversal, it could threaten the entire stock.

ACKNOWLEDGMENTS

This study was supported by a grant to the first author by the Post-Graduate Federal Agency CAPES/MEC, Brazil. Thanks are due to the National Council for Scientific and

Technological Development of the Ministry of Science and Technology of Brazil (CNPq/ MCT) for the Research grants offered to the second and third authors.

REFERENCES

Allen, G. R., Steene, R. and Allen, M. (1998). *A guide to angelfishes & butterflyfishes*. Odyssey Publishing, Tropical Reef Research, Washington, 250p.

Barros, N. H. C. (2012). Estratégia reprodutiva do peixe hermafrodita mussum, *Synbranchus marmoratus* em um açude do Rio Grande do Norte, Brasil. Masters Dissertation, Universidade Federal di Rio Grande do Norte, Natal, Brazil. 84p.

Billard, R., Fostier, A., Weil, C. and Breton, B. (1982). Endocrine control of spermatogenesis in teleost fish. *Canadian Journal of Fisheries and Aquatic Sciences*, 39, 65-79.

Brandão, C. A., Valentim, A. F. M. and Pellegrini-Caramaschi, E. (2003). Ovary maturation stages and oocyte features in three species of the Neotropical Fish *Hemiodus* (Müller, 1842). *Brazilian Archives of Biology and Technology*, 46, 433-441.

Britski, H.A., Silimom, K. Z. S. and Lopes, B. S. (1999). *Peixes do Pantanal: Manual de identificação*. H. A. Britski (Ed), Brasília: EMBRAPA, p184.

Carruth, L. L. (2000). Freshwater Cichlid is a Protogynous Sequential Hermaphrodite. *Copeia*, 1, 71-82.

Chan, S. T. H. and Yeung, W. S. B. (1983). Sex control and sex reversal in fish under natural conditions. In: Hoar, W.S., Randall, D.J. and Donaldson, E.M. (Eds.), *Fish Physiology*, Academic Press, London. 9, 171-222.

Chellappa, S., Bueno, R. M. X., Chellappa, T., Chellappa, N. T. and Val, V. M. F. A. (2009). Reproductive seasonality of the fish fauna and limnoecology of semi-arid Brazilian reservoirs. *Limnologica*, 39, 325- 329.

Devlin, R. H. and Nagahama, Y. (2002). Sex determination and sex differentiation in fish: an overview of genetic, physiological, and environmental influences. *Aquaculture*, 208, 191-364.

Grier, H. J. and Taylor, R. G. (1998). Testicular maturation and regression in the common snook. *Journal of Fish Biology*, 53, 521-542.

Gomiero, L. M., Villares Junior, G. A. and Naous, F. (2009). Reproduction of *Cichla kelberi* (Kullander and Ferreira, 2006) introduced into an artificial lake in southeastern Brazil. *Brazilian Journal of Biology*, 69, 175-183.

Hibiya, T. (1982). *An atlas of fish histology: Normal and pathological features*. Gustav Fischer Verlag, Stuttgart. 147p.

Liem, K. F. (1963). Sex reversal as a natural process in the Synbranchiform fish *Monopterus albus*. *Copeia*, 1963, 303–312.

Liem, K. F. (1968). Geographical and taxonomic variation in the pattern of natural sex reversal in the teleost fish order Synbranchiformes. *Journal of Zoology*, London, 156, 225-238.

Lo Nostro, F. L. (2000). Espermatogénesis, ciclo anual e inducción hormonal de La espermiación en el pez protogínico diándrico, *Synbranchus marmoratus* Bloch, 1795 (Teleostei, Synbrnachidae). Doctoral Thesis, Facultad de Ciências Exactas y Naturales, Universidade de Buenos Aires, Buenos Aires. 138p.

Lo Nostro, F. L. and Guerrero, G. A. (1996). Presence of primary and secondary males in a population of the protogynous fish *Synbranchus marmoratus*. *Journal of Fish Biology*, 49, 788-800.

Montenegro, L. A., Damasceno, D. N. F., Almeida, R. G. and Chellappa, S. (2011). Biologia alimentar do mussum, *Synbranchus marmoratus* (Bloch, 1795) (Osteichthyes: Synbranchidae) no açude Marechal Dutra localizado no semi-árido brasileiro. *Biota Amazônia*, 1 (2), 52- 60.

Montenegro, L. A., Silva, N. B., Nascimento, W. S. and Chellappa, S. 2012. Anatomy and Histology of the digestive tract and feeding habits of the marbled swamp eel *Synbranchus marmoratus*. *Animal Biology Journal*, 3 (3): 127-143.

Moyer, J.T. (1990). Social and reproductive behavior of *Chaetodontoplus mesoleucus* (Pomacanthidae) at Bantayan Island, Philippines, with notes on pomacanthid relationships. *Japanese Journal of Ichthyology*, 36, 459-467.

Munday, P.L., Buston, P.M. and Warner, R. R. (2006). Diversity and flexibility of sex-change strategies in animals. *Trends in Ecology and Evolution*, 21, 89-95.

Moreno, C.A., Urzúa, R. and Bahamonde, N. (1977). Breeding season, sexual rate and fecundity of *Basilichthys australis* Eigenmann 1927, from Maipo River, Chile. (Atherinidae, Pisces). *Studies on Neotropical Fauna Environment*, 12, 217-223.

Murata, K., Sugiyama, H., Yasumasu, S., Iuchi, I., Yasumasu, I. and Yamagami, K. (1997). Cloning of cDNA and estrogen-induced hepatic gene expression for choriogenin H, a precursor protein of the fish egg envelope (chorion). *Proceedings of the National Academy of Sciences,* USA, 94, 2050-2055.

Pudney, J. (1995). Spermatogenesis in nonmammalian vertebrates. *Microscopy Research and Technique*, 32, 459-497.

Reinboth, R. (1983). The peculiarities of gonad transformation in teleosts. *Differentiation*, 23 (Suppl.), 82-86.

Reinboth, R. and Brusle-Sicard, S. (1997). Histological and ultrastructural studies on the effects of hCG on sex inversion in the protogynous teleost *Coris julis*. *Journal of Fish Biology*, **51,** 738-749.

Sadovy, Y. and Shapiro, D. Y. (1987). Criteria for the diagnosis of hermaphroditism in fishes. *Copeia,* 1987, 136-156.

Shapiro, D. Y. (1987). Differentiation and evolution of sex change in fishes. *Biology Science*, 37, 490-497.

Smith M, P., Sansom, I. J. and Repetski, J. E. (1996). Histology of the first fish. *Nature*, 380, 702-704.

Thresher, R. E. (1984). *Reproduction in Reef Fishes*. T.F.H. Publ., Neptune City, 398p.

Wallace, R. A. and Selman, K. (1981). Cellular and Dynamic Aspects of Oocytes Growth in Teleosts. *American Zoologist*, 21, 325-343.

West, G. (1990). Methods of assessing ovarian development in fishes: a review. *Australian Journal of Marine and Freshwater Research*, 41, 199-222.

Yoshida, H. O. (1964). Skipjack tuna spawning in the Marquesas Islands and Tuamotu Archipelago. *Fishery Bulletin*, 65, 479-488.

In: Biology of Semiarid Tropical Fish
Editors: J. R. Gomes and S. Chellappa
ISBN: 978-1-63117-067-6

Chapter 4

MATE SELECTION AND REPRODUCTIVE SUCCESS IN THE AMAZONIAN ANGELFISH, *PTEROPHYLLUM SCALARE* (*OSTEICHTHYES: CICHLIDAE*)

***Sathyabama Chellappa*[1]**, *Maria do Socorro Ribeiro Freire Cacho*[2] *and Gilson Luiz Volpato*[3]**

[1]Post-Graduate Programme in Psychobiology, Centro de Biociências, UFRN, Campus Universitário, Natal, RN, Brazil
[2] Departmento de Ciências Animais, Universidade Federal Rural do Semi-Árido, Bairro Pres. Costa e Silva, Mossoró, Rio Grande do Norte
[3]Laboratory of Animal Physiology and Behaviour, Research Centre on Animal Welfare – RECAW, Department of Physiology, IB, CAUNESP, UNESP, Botucatu, SP, Brazil

ABSTRACT

Mate selection and mating success in the Amazonian angelfish, *Pterophyllum scalare* were experimentally investigated in the laboratory. This species has a breeding system where male body size and substrate type could be decisive for mate selection by females. The aims of this study were to identify the male traits contributing to reproductive success, and to evaluate preferences for each male trait and their relative importance for the females. Different types of Amazonian aquatic vegetation were provided as potential spawning substrates, which were selected and defended by males with varying levels of aggression. Laboratory observations revealed that large sized males were aggressive, dominant and involved in agonistic interactions attacking intermediate and small sized males. Body size and dominance of males were related to space use and proximity to potential spawning sites. The female assessed the male based on two important traits: the type of spawning substrate and the body size. The quality of the territory is the most important trait for females to choose the mate. The larger males defending low quality spawning substrate (aquatic plants with narrow and small leaves) were discriminated against smaller males defending good quality spawning substrate

* E-mail of the corresponding author: chellappa.sathyabama63@gmail.com.

(aquatic plants with broad leaves, *Anubia* sp. and *Echinodorus amazonensis*). Female preference for substrate type was independent of male body size. When given a choice between males matched for size but differing in the type of substrates defended, females preferred the male defending good quality spawning substrate. When substrate type was held constant females preferred larger males. Though both traits may be strong predictors of the male's competitive ability, the substrate type was the main determinant of male reproductive success.

Keywords: Male mating success; female mate preference; *Pterophyllum scalare*; Amazonian cichlid fish

INTRODUCTION

Intra-sexual competition between males and inter-sexual mate choice by females may have complementary or antagonistic effects on the evolution of sexually-selected traits. Sexual selection is primarily driven by interactions between individuals of the same sex (intra-sexual competition for access to mates) and interactions between individuals of opposite sexes (inter-sexual mate choice) (Andersson, 1994; Hudman and Gotelli, 2007).

Information on the phenotypic correlates of male reproductive success can provide important insights into the operation of sexual selection, and the nature of evolutionary forces on phenotypic traits such as male body size. Sexual selection theory predicts strong selection for traits that increase reproductive success in both sexes. In males, mate acquisition is often the most severe limitation on reproductive success and, consequently, male strategies to obtain matings with more than one female are common (Say et al. 2003; Uller and Olsson 2008). The role of body size in enhancing male mating success provides an ideal system to investigate the nature of adaptation. Both body size and mating success are amenable to quantification in the field, and the latter variable is so intimately tied to fitness (i.e., individual reproductive success) that there is no need to assume fitness relevance, as is the case for most other traits. Empirical work has revealed that larger males experience higher mating success in a wide variety of organisms, but interpretations of the functional basis for such an association generally are inferential only (Shine and Mason, 2005).

Difference in body size among fish conspecifics is a remarkable phenomenon associated with many biological processes. Such heterogeneous growth has genetic causes, but is greatly affected by population factors (Volpato and Fernandes, 1994). As a consequence, several processes are affected by fish body size, mainly mediated by aggression. Larger fish are usually dominant (Turner, 1994; Chellappa et al., 1999a), although aggression may surpass the body size effect on dominance (Moretz, 2003). Isolated fish, owners of territories and previous dominants usually win combats because aggression is increased in such conditions (Beaugrand et al., 1996; Cacho et al., 1999; Whiteman and Côté, 2004). In territorial fish species the body size determines social hierarchy, which in turn distributes environmental resources accordingly. Larger fish usually get priority to territory, food (Whiteman and Côté, 2004), shelters (Orpwood et al., 2003), spawning sites (Itzkowitz et al., 1998) and mates (Karino, 1995).

In many teleost fishes the male breeding behaviour consists of exclusive occupation of a breeding site where the females are courted (Breder and Rosen, 1966; Huntingford, 1993; Chellappa et al., 1999b). Among fishes the opportunity for a male to receive and care for eggs is related to the possession of a territory suitable for egg deposition. This idea has also been used to predict the levels of resource availability that could lead to social hierarchy.

The male characteristics, such as, relative body size and aggression, coupled with the possession of a good quality territory, could cause differences in individual mate success (Arnold, 1983). Body size mediated competition may have different meanings depending on whether the reproductive context is involved. A male-male competition may be important for resources like hiding places, food and space. In a reproductive context, mating condition may modulate the importance of some cues in the environment (Volpato et al., 2004).

During mating, both males and females attempt to balance the trade-off between the costs of mating and the fitness benefits gained from employing a certain strategy, and the balance is not necessarily complementary. Teasing apart the influences of intrasexual and intersexual components of sexual selection is difficult, however, males and females interact and respond simultaneously to one another. Experimental studies that allow for separate measures of male–male competition and female choice in the same individuals can overcome this limitation (Hudman and Gotelli, 2007).

The graceful angelfish, *Pterophyllum scalare* originates from the Amazonian basin, prefers shallow and calm waters with aquatic vegetation and forms small schools. Aggressiveness and territoriality are expressed with great ferocity by males during the reproductive phase, when they guard their nests and young. Throughout the breeding period these fish demonstrate aggressive behaviors such as threats, attacks and bites (Chellappa et al., 1999a; 1999b) and is a valuable ornamental fish, very popular all over the world.

On the basis of these considerations, the present study investigated the traits contributing to female mate preference and male mating success in the angelfish, *Pterophyllum scalare*, a cichlid from the Amazonian basin. This species was selected, since it is an important fish model for studying aggressive behaviour and territoriality (Chellappa et al., 1999a, 1999b; Cacho et al., 2007) and both aspects are related to the proposed investigation.

The following questions were addressed in the present study:

- During intra-sexual competition between males, how does body size relate to dominance, proximity to good quality spawning substrate and mates?
- What are the traits preferred by females during mate choice?
- What are the strong predictors of the male's competitive ability which contribute to its reproductive success?

MATERIALS AND METHODS

This work verified the following phases during a reproductive cycle: i) intra-sexual competition between males; 2) selection of substrate and establishment of spawning territory; 3) inter-sexual mate choice of females; 4) courtship and mating and 5) parental care.

Acquisition and Maintenance of Fish

The angelfish, *Pterophyllum scalare* used in this study (N = 60; 45 males and 15 females) originated from the tributaries of the River Tapajós of the Amazon Basin, Brazil. All of them belonged to the zero year class and the sample size used was sufficiently large to allow behavioural observations without overexploiting the natural fish population. The males were identified on the basis of the presence of genital papillae and were housed separately from the females in two separate 1000 L cement tanks. This precaution was necessary to avoid previous recognition effect. The tanks were provided with constant aeration and the fish were fed daily *ad libitum* with commercially available diet for cichlids and newly hatched *Artemia salina*. The concentration of dissolved oxygen was maintained at 8.0 mgL^{-1}, pH values from 6.8 to 7.0 and water temperature around 27^0C. Precautions were taken to minimize disturbance to the fish and the natural photoperiod of 06:00 to 18:00h was maintained in the Ichthyology Laboratory of the Department of Oceanography and Limnology, Federal University of Rio Grande do Norte, Natal, Brazil. The fish were left undisturbed for three weeks prior to the beginning of the experiments, to allow them to acclimate to the laboratory conditions. Before the experiments began, the length of all fish, from the tip of the snout to the end of the caudal fin, was measured to the nearest 0.1cm. All behavioural observations were carried out from October, 2006 to January, 2007 during the most conducive period for reproduction, when spawning occurs in the natural environment.

Types of Spawning Substrates

Observations in the laboratory were carried out under semi-natural conditions, similar to those encountered in nature. In their natural habitat, these fish live along river margins in clear water with dense aquatic vegetation, and are usually found admist the submerged parts of these plants.

Fifteen glass aquaria (70x 40x40 cm) were set up in the laboratory with crushed stone and sand bottom, planted with different species of Amazonian aquatic vegetation (*Echinodorus amazonensis* and *Anubia* sp. with broad leaves and *Valisneria gigas* and *Elodea* sp with long narrow leaves). These are native plants found in the mid to lower region of the Amazon River basin and were selected as potential spawning site options. *E. amazonensis* has lanceolate and arched leaves which are 2–3 cm wide whereas *V. gigas* has narrow leaves of 1 cm width.

Intra-Sexual Competition between Males for Spawning Sites

To investigate the effect of body size of males on aggression and dominance, the experimental design consisted of focal observations of fifteen groups of the angelfish (N = 45 males). Each group consisted of three different sized sexually mature males, one large (13.40 ± 0.96cm), one medium sized (11.40 ± 0.65cm) and one small (8.70 ± 0.34cm). All males were introduced into aquaria at the same time, and three days were allowed to settle down, establish dominance and select spawning sites. At the end of this period, types of aggressive interactions were identified and counted. Behavioural observations were carried out for two consecutive days, for a period of 120 minutes each day (09:00 to 10:00 and 15:00 to 16:00h). A total of 240 minutes of observations on interactions were focused on various stages of fights, such as, approximation, permanence, threat, attacks, bites, following and escape (Chellappa et al. 1999a and 1999b).

Selection of Substrate and Establishment of Spawning Territory

Intra-sexual competition between males for proximity to aquatic plants was observed. Selection was confirmed when the dominant male remained close to a specific substrate and defended this territory for at least 60% of the total observation period. Male positions were determined in relation to the potential spawning sites.

Female Mate Preference

After the establishment of territories, one sexually mature female was concomitantly introduced into each aquarium containing the three males. Behavioural observations were carried out for two consecutive days after the introduction of the females, for a period of 60 minutes, twice a day (09:00 to 10:00 and 15:00 to 16:00h).

Observations of interactions between males were focused on fights (attacks and bites). Female mate preference was identified by maximum time spent (at least 60% of the total observation period) with a male near a specific spawning substrate. Male dominance was confirmed by proximity to spawning site and maximum time spent by the female near the selected male's territory.

Female Preference for Substrate Quality

This experiment examined whether female preference for substrate quality was independent of male body size effects: 1) males of equal size but differing in the type of substrates defended (aquatic plants with broad leaves or with narrow leaves); 2) males of three different sizes defending similar substrates (vegetation type was held constant). Female preference was observed in aquaria as described earlier. Different females were used for the experiment and the control. Fifteen replicates were conducted in the experimental condition (with males) and in the control condition (without males).

The Relative Importance of Male Size and Substrate Quality

This experiment tested the relative impacts of substrate quality and male size on female preferences, by giving females a choice between a large male with a low-quality substrate (aquatic plant with narrow leaves) and a small male with a high-quality substrate (aquatic plant with wide leaves). Experimental aquaria were set up as described earlier and fifteen replicates were run. The eggs laid by each pair were counted and reproduction was monitored till the eggs hatched out and till the end of the reproductive cycle.

Statistical Analyses

Frequency data were transformed (square root) for parametric analyses. Comparison of male body size considered mean values and standard deviations of 3 sizes (large, intermediate and small) and 2 conditions (in the absence and presence of female). A 3 x 2 model of ANOVA was used to analyse male fish behaviour before and after female introduction. When statistical significance was detected ($\alpha = 0.05$), individual means were compared by post-hoc Tukey.

RESULTS

Male Body Size and Level of Aggression

The results show that when males initiate attacks the frequency of aggression was significantly different among males of the three different body sizes. The level of aggression increased proportionately with the body size of male fish (Fisher/Snedecor, $F=58.08$; $p<0.01$). Prior to the introduction of females, large males were aggressive, dominant and involved in agonistic interactions attacking both intermediate and small sized fishes (Fisher/Snedecor, $F=163.46$; $p<0.05$). The intermediate sized fish concentrated attacks on small fish. Thus small sized fishes were recipients of attacks from both large and intermediate sized males. After females were introduced into the aquaria, the frequency of attacks increased only in large fish ($p<0.01$), with no significant changes in other males (Figure 1).

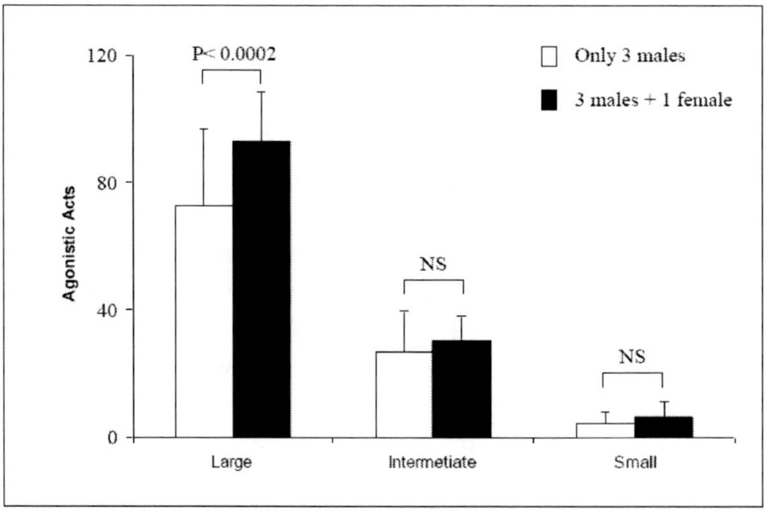

Figure 1. Fights among angelfish males in relation to body size and group composition. $F=163.46$; $p<0.05$.

Space Use and Male Body Size

Male body size and/or dominance were related to space use prior to the introduction of females. Consequently, small fish occupied a reduced space according to their relative body size in comparison to large males (Figure 2). After the introduction of females, males changed the space occupation; however a clear response was identified specifically in large males. The intermediate size fish remained nearer to large fish in relation to small fish, which in turn maintained relatively large distance from the other two males. Space occupation was different for males in each condition, being larger for the large fish, comparatively small for the intermediate sized fish and much smaller for the small sized fish (Figure 2).

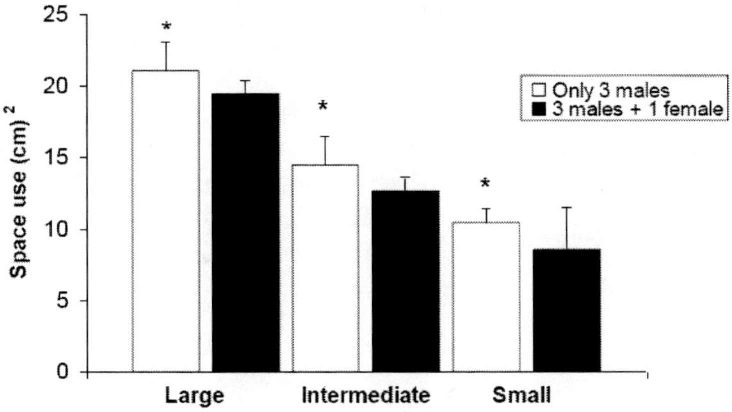

Figure 2. Space use among angelfish males in relation to body size and group composition. F=359.65; p<0.05.

Proximity to Aquatic Plants

Large males were nearer to wide leaves of *E. amazonensis* and intermediate size fish remained close to long narrow leaves of *V. gigas*, whereas small fish remained far away from wide leaf plants and fairly distant from narrow leaf plants. The correlation between aggression of males and proximity to aquatic plants was significant (Person, r= 0.59, p<0.05). After the introduction of female, both large and intermediate size fishes were closer to the aquatic plants. Therefore, the distance between these fishes increased in relation to the small fish (Figure 3).

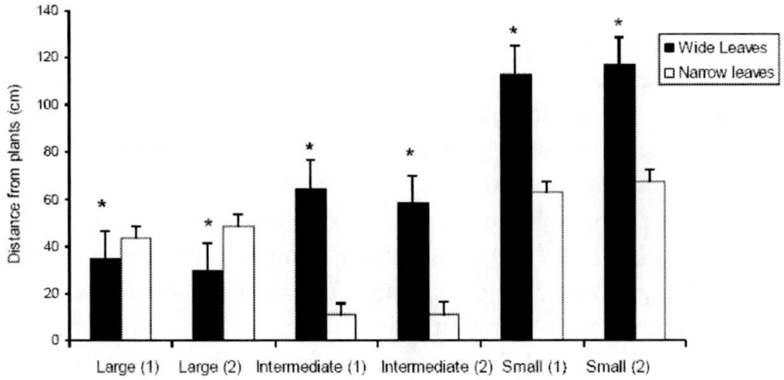

Figure 3. Proximity to aquatic plants among angelfish males in relation to body size and group composition. F= 14.20, p<0.05 1 = 3 males 2 = 3 males + 1 female.

Partner Selection and Mating Success

The female when introduced to a group of three males, ignored the intermediate and small sized fish and selected the big aggressive male in possession of potential spawning site. Consequently, the female spent more time near large male close to wide leaved aquatic plants. The correlation between proximity to large male and the time spent by the female was significant (Person, $r= 0.67$, $p<0.05$).

The female spawned always with the largest male of the group of three (ANOVA, t-test, $t=15.58$; $p<0.050$). The adhesive eggs were deposited on the surface of wide leaves of the aquatic plants. The angelfish is a partial spawner and the number of eggs per batch ranged from 130 to 210, with a mean value of 176 ± 32.09 ($\pm$ SD).

The pair guarded the eggs and a period of bi-parental care was observed before the hatching of the eggs in 48 to 72 hours. The wrigglers hang suspended from the leaves by mucous threads. The intermediate and small sized males did not participate in spawning.

The duration of the reproductive cycle ranged from 2 to 6 days, with an average value of 3.80 ± 1.79 days.

Discussion

In group-living animals, pair-wise encounters occur and sometimes individuals can be arranged to produce a linear ranking system or dominance hierarchy which summarises the pattern of agonistic relationships within the group. Dominant animals often get priority of access to limiting resources, shelters, food and mates (Huntingford and Chellappa, 2011).

The present study showed that social dominance is a function of body size in angelfish, which gives the dominant male fish priority to spawning sites and females. Moreover, the preferred type of spawning site is related to the size of the leaves of aquatic vegetation, where the eggs are laid. Space use was also related to body size and aggression resulting in differential occupation among males during reproductive competitions.

Dominance associated with body size has been widely reported in fish (Turner and Huntingford, 1986; Turner, 1994), including in the Amazonian angelfish (Chellappa et al., 1999a). It was observed that aggression is a characteristic of dominant fish which also gives access to mates.

The dominant fish choose the best place to establish its territory, which is determined by a combination of different cues from the environment (Huntingford, 1993). In the present study, the environment was homogeneous in terms of light and temperature and the only difference was due to the presence of two different types of aquatic plants. There is a clear association of dominance with potential spawning sites and the fish preference for specific aquatic plants. Large dominant fish preferred the plant *E. amazonensis* with wide leaves to lay eggs. These plants are native to the Amazon River basin and are with wide leaves ideal for even deposition of the egg mass during spawning. Intermediate size fish stayed closer to the plant *V. gigas* with long narrow leaves. It could be argued that the intermediate-sized fish could have selected the plants with long narrow leaves, not because it was the second best spawning site option available, but because it was a good place for hiding from aggression of the dominant male. However, the large dominant fish had the priority to choose the better option, whereas the intermediate fish had a restricted selection and the small fish was left with no selective preference. As the larger fish could, in fact, choose the preferred aquatic plant, where eggs were laid latter on, the adequacy of the leaf surface to support eggs could be considered as a trait of evolutionary significance to this species.

The striking difference in relative size among the Amazonian male angelfish is the keynote for dominance, space use and mates. Only dominant fish reproduced. In this context, aggression was important to secure these environmental resources (Torricelli et al., 1993). Thus, in the Amazonian angelfish aggression is a mean to assure mating success wherein large body size of the males influenced aggression, dominance, proximity to preferred egg laying sites and mating success.

A trend for larger males to obtain a disproportionately high number of matings, as occurs in many animal populations, typically is attributed either to female choice or success in male-male rivalry. If large dominant males tend to have high mating success, then the ultimate result may be the evolution of large male body size (Maynard-Smith and Brown 1986) and male-biased sexual size dimorphism. In mating systems where males compete physically and where larger body size confers a significant advantage in male-male competition, the actual selective force for larger body size in males may relate to possession of good quality spawning territories.

ACKNOWLEDGMENTS

This study was supported by the National Council for Scientific Research and Technological development of the Ministry of Science and Technology (CNPq-MCT), Brazil, in the form of Research grants during the study period.

REFERENCES

Andersson, M. (1994). *Sexual selection*. Princeton University Press, Princeton, NJ. Arnold, S.J. (1983). Sexual selection, natural selection and quality advertisement. *Biological Journal of the Linnean Society*, 17, 375-393.

Beaugrand, J.P., Payette, D. and Goulet, C. (1996). Conflict outcome in male green swordtail fish *Xiphophorus helleri*: interaction of body size, prior dominance/subordination experience and prior residency. *Behaviour*, 133, 303-319.

Breder, C.M., Jr. and Rosen, D.E. (1966). *Modes of Reproduction in Fishes*. American Museum of Natural History Press, New York.

Cacho, M.S.R.F., Yamamoto M.E. and Chellappa, S. (1999). Comportamento reprodutivo do acará bandeira, *Pterophyllum scalare* (Osteichthyes, Cichlidae). *Revista Brasileira de Zoologia*, 16 (1), 653-664.

Cacho, M.S.R.F., Yamamoto M.E. and Chellappa, S. (2007). Mating system of the Amazonian cichlid angel fish, *Pterophyllum scalare*. *Brazilian Journal of Biology*, 67 (1): 161 – 165.

Chellappa, S., Yamamoto, M.E., Cacho, M.S.R.F. and Huntingford, F.A. (1999a). Prior residence, body size and the dynamics of territorial disputes between male freshwater angelfish. *Journal of Fish Biology*, 55, 1163-1170.

Chellappa, S., Yamamoto, M.E. and Cacho, M.S.R.F. (1999b). Reproductive behaviour and ecology of two species of cichlid fishes. p.113-126. In: Vall A.L., Vall, V.M. (Eds.), *Biology of Tropical Fishes*, INPA, Manaus.

Hudman, S.P. and Gotelli, N.J. (2007). Intra and intersexual selection on male body size are complimentary in the fathead minnow (*Pimephales promelas*). *Behaviour*, 144, 1065-1086.

Huntingford F.A. (1993). Behaviour, Ecology and Teleost Fishes. In: Huntingford, F.A., Torricelli, P. (Eds.), *Behavioural Ecology of Fishes,* Harwood Academic Publishers, Chur.

Huntingford F.A. and Chellappa, S. (2011). Agressão. p. 157–173. In: M. E. Yamamoto and G. L. Volpato (Eds.), *Comportamento Animal*. Second edition, EDUFRN, Natal, Brazil. 342p.

Itzkowitz, M., Vollmer, G. and Rios-Cardenas, O. (1998). Competition for breeding sites between monogamous pairs of convict cichlids (*Cichlasoma nigrofasciatum*): asymmetries in size and prior residence. *Behaviour*, 135, 261-267.

Karino, K. (1995). Male-male competition and female mate choice through courtship display in the territorial damselfish *Stegastes nigricans*. *Ethology*, 100, 126-138.

Maynard-Smith, J. and Brown, R.L. (1986). Competition and body size. *Theoretical Population Biology*, 30, 166-179.

Moretz, J.A. (2003). Aggression and RHP in the northern swordtail fish, *Xiphophorus cortezi*: The relationship between size and context dynamics in male-male competition. *Ethology*, 109, 995-1008.

Orpwood, J.E., Griffiths, S.W. and Armstrong, J.D. (2003). Effects of body size on sympatric shelter in over-wintering salmonids. *Journal of Fish Biology*, 63 (supp. A), 166-173.

Say, L., Naulty, F. and Hayden, T. J. (2003). Genetic and behavioural estimates of reproductive skew in male fallow deer. *Molecular Ecology*, 12, 2793–2800.

Shine, R. and Mason, R.T. (2005). Does large body size in males evolve to facilitate forcible insemination? A study on garter snakes. *Evolution,* 59(11), 2426–2432.

Torricelli, P., Lugli, M. and Bobbio, L. (1993). The importance of male-male competition and sexually selected dimorphic traits for male reproductive sucess. In: Huntingford, F. A., Torricelli, P. (Eds.), *Behavioural Ecology of Fishes*, Harwood Academic Publishers, Chur.

Turner, G.F. (1994). The fighting tactics of male mouthbrooding cichlids: the effects of size and residency. *Animal Behaviour*, 47, 655-662.

Turner, G.F. and Huntingford, F.A. (1986). A problem for Game Theory analysis: assessment and intention in male mouth brooder contests. *Animal Behaviour*, 34, 961-970.

Uller, T. and Olsson, M. (2008). Multiple paternity in reptiles: patterns and processes. *Molecular Ecology*, 17, 2566-2580.

Volpato, G.L., Duarte, C.R.A. and Luchiari, A.C. (2004). Environmental color affects Nile tilapia reproduction. *Brazilian Journal of Medical and Biological Reserch*, 37, 479-483.

Volpato, G.L. and Fernandes, M.O. (1994). Social control of growth in fish. *Brazilian Journal of Medical and Biological Reserch*, 27, 797-810.

Whiteman, E.A. and Côté, I.M. (2004). Dominance hierarquies in group-living cleaning gobies: causes and foraging consequences. *Animal Behaviour*, 67, 239-247.

In: Biology of Semiarid Tropical Fish
Editors: J. R. Gomes and S. Chellappa
ISBN: 978-1-63117-067-6

Chapter 5

DYNAMICS OF TERRITORIAL BEHAVIOUR AND GONAD DEVELOPMENT IN THE HYBRID RED TILAPIA, *OREOCHROMIS NILOTICUS* X *O. MOSSAMBICUS* (*OSTEICHTHYES: CICHLIDAE*)

Sathyabama Chellappa[1]*, Ana Patrícia Targino de Medeiros[1], Maria do Socorro Ribeiro Freire Cacho[2] and Gilson Luiz Volpato[3]

[1]Postgraduate Program in Psychobiology, Department of Physiology, Universidade Federal do Rio Grande do Norte, Av. Salgado Filho, Lagoa Nova, Natal, Rio Grande do Norte, Brazil

[2] Departmento de Ciências Animais, Universidade Federal Rural do Semi-Árido, Bairro Pres. Costa e Silva, Mossoró, Rio Grande do Norte

[3]Laboratory of Animal Physiology and Behaviour, Research Centre on Animal Welfare – RECAW, Department of Physiology, IB, CAUNESP, UNESP, Botucatu, SP, Brazil

ABSTRACT

The hybrid red tilapia, *Oreochromis niloticus* x *O. mossambicus*, is utilized in semi-intensive fish culture practices in Northeastrn Brazil, and this paper reports on the reproductive behaviour and gonadal development of this hybrid. Mature males and females of hybrid red tilapia were observed under laboratory conditions during their reproductive phase, to verify behaviour related to dominance during agonistic interactions, nest building, courtship, mate selection, mating and parental care. Histological examinations of ovaries and testes were carried out in different developmental stages to identify maturity stages and to determine the gonadosomatic index, fecundity and type of spawning. Aggression and dominance exhibited by males during agonistic interactions are influenced by their body size. The gonadosomatic indices of males show positive correlations with their dominance level. Large and aggressive males in possession of spawning sites were preferred by females during mate selection. The ovaries showed immature (phase I and II), initial maturing (phase II and

* E-mail of the corresponding author: chellappa.sathyabama63@gmail.com.

III), final maturing (phase IV and V), mature (phase I to V) and partially spent stages. The testes showed immature, maturing, mature and partially spent stages during development. Ovaries of mature females presented synchronic oocyte development indicating multiple spawning, with mean batch fecundity of 1.213 mature oocytes. Parental care was exhibited by the pair after mating and males were more aggressive than females during egg and larval protection. Effective participation on the part of females during parental care phase resulted in higher survival rate of larvae.

Keywords: Reproduction, agonistic interactions, gonadosomatic index, parental care, hybrid red tilapia

INTRODUCTION

The cichlid fishes have a wide geographical distribution and they are distributed in Africa, South and Central America, Southeastern part of North America (South of Texas and México), Madagascar, India, Sri Lanka, Syria, Israel and Iran (Kullander, 1998). The family Cichlidae comprises of 1600 valid species. Tilapia is the common name given for nearly a hundred species of cichlid fish, which inhabit a variety of fresh water habitats. Breeding of tilapia adapted to temperate climates and saline environments was designed to exploit the ease of producing inter-specific hybrids among tilapias. Triploids have been produced in *Oreochromis niloticus*, *O. mossambicus* and *O. aureus*, and tetraploids have been induced in *Oreochromis niloticus* and *O. mossambicus*. In the case of polyploid hybrids, it is expected that gonadal development may be suppressed and consequently additional energy may be channeled into increased somatic growth (Keenleyside, 1991).

The substrate guarding cichlids are generally monogamous and the pair protects their eggs and larvae in a breeding territory as in *Cichla monoculus* (Chellappa et al., 2003) or are polygamous. Similarly, the mouth-brooding species are either monogamous as in *Sarotherodon galilaeus* or polygamous where either males or females carry their eggs and larvae in their mouths (Barlow, 1991). The evolution of mating systems in cichlid fishes has proceeded from monogamy with biparental care to polygamy with maternal care, with a number of variations on these two themes (Cacho et al., 2007).

The long period of parental care of the young after spawning characterizes the family Cichlidae. During this period, which lasts for about three weeks, the offspring is protected by one or both parents, an interesting feature which attracts the attention of investigators and aquarists of fish reproductive behavior and ecology.

Biparental care effectively removes both parents from further reproduction during care (Chellappa et al., 1999a). It may be expected that each parent be tempted to desert the offspring in order to invest in additional matings. Maternal mouth brooding probably provides the opportunity for males to abandon their mates. Since female invests energetically more in egg production than does male in sperm, consequently male is expected to desert more often than female. Monogamous male cichlids tend to desert and females remain and protect with the young (Cacho et al., 1999; Cacho et al., 2007).

Substrate brooders are species in which the parents maintain their brood on or close to the substrate from spawning until the young become independent. On the other hand, mouth brooders are species in which the parents carry their offspring in their mouth until the young become independent.

The protective role begins as soon as the eggs are shed and ends when the link between parents and offspring is broken as the fish larvae begins independent life. The parents attack and chase away any intruders that approach their brood. The spawning location can influence the success of parental guarding (Cacho et al., 1999; Cacho et al., 2007).

In cichlids, competition among males is more evident as a sexual selection process that influences the reproductive success of individuals (Cacho et al., 2006). During the breeding season, males establish a territory and compete for females directly through mate defense or indirectly by defending the territory to which the females are attracted by their courtship behaviour (Keenleyside, 1991; Barber et al., 2001; Chellappa et al., 2003). During courtship females select their mates based on various physical attributes, such as body size and coloration and dominant males are generally chosen by females (Ridley, 1995). Social interactions affect several processes of aquaculture interest, such as growth (Volpato and Fernandes, 1994), stress (Fernandes and Volpato, 1993; Alvarenga and Volpato, 1995) and reproduction (Chellappa et al., 1999a, 1999b).

Hybrid red tilapia, *Oreochromis niloticus* x *O. mossambicus* was introduced in the Fish culture Station located in Macaíba, Rio Grande do Norte, Brazil, and is utilized in semi-intensive fish culture practices in Northeastern Brazil (Chellappa, et al., 1996; Chellappa, 2000). Though there is a great potential for the culture of this hybrid, research articles published about this fish are scarce. As such, biological information about this hybrid is important for further studies besides its application in regular fish culture management practices. This paper reports on reproductive behavior involving aggression and dominance during territorial disputes, competition for spawning site and mates, parental care exhibited besides the gonadal development in the hybrid red tilapia

Materials and Methods

Fish Acquisition and Maintenance

Fifty males and thirty females of mature hybrid red tilapia, *Oreochromis niloticus x O. mossambicus* were acquired from the Fish culture Station of the Agriculture College of Universidade Federal do Rio Grande do Norte, Macaíba, Brazil.

All fish originated from a stock of under one year old fishes, mean total length of males was 16.4 cm ± 0.82 cm (maximum of 22.5 cm and minimum of 13.0 cm, ± 0.98 cm), and females measured 12.2 cm ± 0.92 in mean total length (maximum of 14.0 cm and minimum of 10.0 cm, ± 0.55 cm).

In the laboratory male and female fishes were maintained separately in two 1000 L cement storage tanks (1.00 x 1.50 m) for two months prior to the observations. Adequate survival conditions were maintained with water temperatures ranging from 26°C to 28°C, pH between 7.0 and 7.5 and concentration of dissolved oxygen between 7.0 and 8.0 mg L^{-1}. Food consisted of cichlid pellets (38% protein) offered twice daily (early morning and late afternoon), corresponding to 3% of the mean fish body weight.

Behavioral Observations

To investigate the reproductive behavior of males and females, the experimental design consisted of observations of three groups of hybrid red tilapia, where each group consisted of 5 males and 2 females. Each group of fish was placed in separate glass aquaria, each measuring 1. 20 x 80 x 80 cm. Bottom of the aquaria were covered with 3 cm layer of sand and fine gravel.

The fish were observed for 30 consecutive days (1 hour per day). Breeding of randomly-formed pairs was observed during one breeding cycle which lasted for approximately 21 days. The following behaviours were observed: dominance disputes, nest building, courtship, mate selection, mating and parental care till free larval stage. Behaviour units utilized in agonistic interactions were based on descriptions of Huntingford (1993), Rogers (1995) and Chellappa et al. (1999a; 1999b).

Dominance

The experimental design to investigate the effect of male body size on aggression and dominance consisted of behavioral observations during agonistic interactions between males in three groups.

Each group was composed of five mature males of three different body sizes, one large male (mean body length of 21.1 ± 0.65 cm), two intermediate sized males (mean length of 16.8 ± 0.96 cm) and two small males (13.0 ± 0.82 cm). All fish were introduced into separate 500 L glass aquaria at the same time. Three days were allowed for the males to settle down, select adequate sites to construct spawning sites and to establish dominance. Observations on agonistic interactions were focused on fights between males (approximation, threat, attacks, bites, pursuing and escape).

Thirty observational sessions each lasting one hour per day were performed. Dominance between hybrid red males was determined by the difference in body size and their aggressiveness and fishes were classified according to attacks emitted and received.

Nest Building

To investigate nest (spawning site) building behaviour during breeding, twenty males were observed in the absence and in the presence of ten mature females. Two different sized males were placed in each aquarium (fish with mean length of 21.1 cm were considered large and those with mean length of 16.8 cm were considered small). Fish were observed during thirty sessions, each of 1 hour duration.

A female was later introduced into each aquarium and fish were observed for another thirty sessions, each of 1 hour duration. Frequency of sand removal, minimum and maximum time spent in nest digging activities and diameter of nest excavated were recorded.

Courtship and Mate Selection

Observations were continued in order to investigate courtship behaviour. Twenty males and ten females were observed during thirty sessions, each of 1 hour duration. Interactions between males were focused on fights (attacks and bites) and the time that female remained beside each male was recorded.

Female preference of mate was confirmed by maximum time spent with one of the males near a spawning site. Courtship acceptance was confirmed when the female remained at least 60% of the observation time beside one of the males. After mate selection, the males which were not preferred were removed. The pairs were subsequently investigated for mating and parental care.

Parental Care

To identify the behaviour displayed during parental care, the pairs were monitored for 60 minutes each day, from egg to free larval stage.

During parental care the diligence of both parents was recorded, based on the defense and care of eggs and larvae. Offspring care was defined as the time male spent close to the female during intra-buccal care of eggs and of larvae, as well as the total time spent by the female in egg care. A male or female intruder was introduced in the presence of the pair and observations were carried out during fifteen sessions, each of 30 minutes duration.

Aggressive reaction to intruders was observed as a measure for protecting eggs and larvae. The eggs laid by each pair were counted and reproduction was monitored for three days until the eggs hatched out. The larval survival rate was calculated.

Gonadal Maturation

Maturation stages of gonads were determined using histological aspects of 20 mature ovaries and 20 testes. To avoid possible variation in the developmental stage of gonads, histological examinations were carried out on sections from the anterior, middle and posterior regions of gonads which were preserved in Bouin's solution for 24 h and then transferred to 50% ethanol.

Preserved gonads were embedded in paraffin, sectioned at 5 - 10 μm thickness and stained with Hematoxylin and Eosin. Gonadal developmental stages were assessed microscopically with the help of light microscope (Taimin, model TM 800), coupled with a video camera (Kodo Digital).

Gonadosomatic Index (GSI)

Eviscerated body weight (Wt in g) and gonad weight (Wg in g) were registered. The gonadosomatic index (GSI) based on wet weights was calculated using the formula of

Wootton et al. (1978): GSI = (Wg x 100) / Wt , where Wg is weight of ovary (g), Wt is body weight of fish (g).

Type of Spawning and Fecundity

Relative frequency of different sizes of oocytes was estimated by enumerating a sub sample of different sized oocytes of 20 mature ovaries (West, 1990). The diameters of oocytes were measured with an ocular micrometer. The terminology utilized for staging individual oocytes based on their histological appearance was adapted from Palmer et al (1995) and fecundity was estimated by counting all mature oocytes in the ovaries (Vazzoler, 1996).

RESULTS

Behavioral Observations

Behaviors presented by hybrid red tilapia included threat (when the fins were erected and operculums were opened as the fish came closer to the opponent), lateral attack (when the fish bit the lateral body side of the opponent), frontal attack (juxtaposition of the fish mandibles with wave like movement of the caudal region), pursuing (when the fish followed the direction of the opponent, either attacking or not), escape (when the fish avoided the site of confrontation), and submission (when the fish maintained the head elevated, the fins close to the body and remained motionless).

Dominance

Dominance in males was determined according to attacks emitted and received. The large fishes emitted more aggression and were considered dominant (α males). Intermediate sized fish received aggression from the α male bur not from the small ones and as such, were considered second dominant (β males).

Small fishes received attacks from both large and intermediate fishes and were considered submissive (sub-β males) (Figure 1). Significant differences were observed regarding aggression during the interactions α-β / β-sub-β, α-β / sub-β- sub-β and α –β/ β –β ($F= 10.74$; $p <0.01$).

Gonadosomatic Index

In the red hybrid, the gonadosomatic index showed a significant correlation with dominance level (Pearson, $r = 0.13$; $p = 0.05$). The dominant fish had higher indices than those of subdominants and submissives. Mean GSI values of the dominants varied from 0.5 to 0.8, with a mean of 0.6. For the subdominant fish, the values ranged from 0.2 to 0.7 with a mean of 0.5, while submissive fish values varied from 0.2 to 0.6 with a mean of 0.4 (Figure 2).

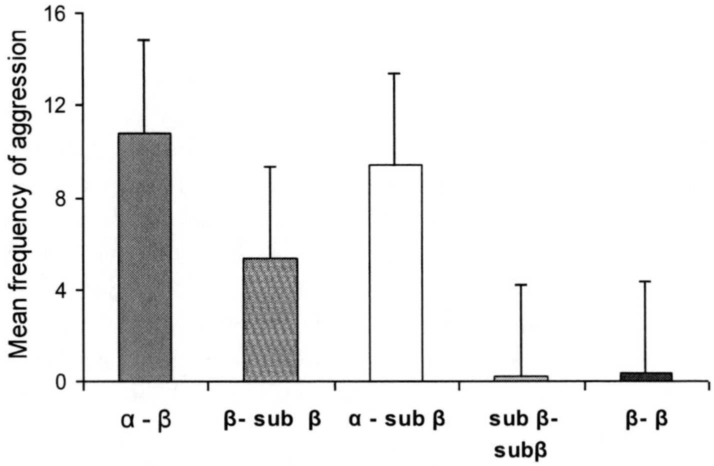

Figure 1. Mean frequency of aggression and dominance between males of hybrid red tilapia, *Orechromis niloticus* x *O. mossambicus* (Student t test: $t= 10.74$; $p< 0.05$).

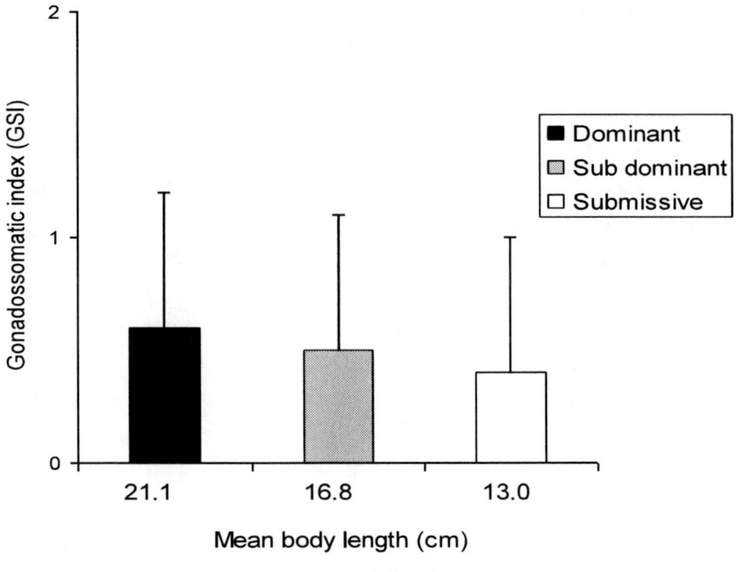

Figure 2. Gonadosomatic index, mean length and level of dominance in males of hybrid red tilapia, *Oreochromis niloticus x O. mossambicus* (Pearson, $r= 0.13$; $p>0.05$).

Nest Building

During nest building, the frequency of sand removal in the absence of a mature female was 18.1 times per minute, whereas in the presence of a mature female it increased to 20 times per minute. The minimum time spent on nest building was 30 minutes and the maximum was 12 hours. Mean nest diameter varied from 12.6 to 21.5 cm, with mean depth of 4.9 cm. Males initiated nest building regardless of the presence of females; however, this activity was intensified in their presence.

Courtship and Mating

In the hybrid red tilapia, the onset of courtship led to the pair formation, which lasted from one to two days. During courtship the male swam in circles (around the female biting her frequently near the tail fin), and removed sand from the nest with its mouth and deposited it nearby. The females remained for a longer time with males who exhibited higher frequency of sand removal and swam more frequently around her. The correlation between the length of time that a female remained with a male and male courtship behaviour was significant (for sand removal: Pearson, r – 0.46; p = 0.00; for swimming in circles: Pearson, r = 0.36; p =0.00). Females remained with larger and more aggressive males for a significantly longer time than with smaller and less aggressive individuals, showing preference for the former, (Student's t-test: t = 36.27; p = 0.00). Courtship behavior of males during the permanence of females were: removal of sand (Pearson, r= 0,46; p= 0,00) and swimming in circles (Pearson, r= 0,36; p= 0,00).

During mating, the sexually mature pair displayed their prominent genital papilla, with the male exhibiting frequent body tremors. At spawning, the female laid the eggs in the nest and the male released semen, fertilizing them. The male also liberated semen into the buccal cavity of the female. The female approached the male genital papilla in order to receive the semen into her mouth. She then gathered the eggs for intra-oral incubation, where larvae eclosion and protection occurred. During the incubation period the female interrupted feeding. On average, 200 eggs were collected, which hatched in 72 hours. The reproductive cycle of hybrid red tilapia had a mean duration of 28 days.

The intra-buccal care provided by the females had a mean duration of 15 days, after which the larvae became free swimming. They returned to the mouth of the female during presence of intruders and remained there until there was no more predation risk. The female collected the larvae for a period of 3 days after which she lost interest in the offspring. In the intra-bucal larval phase, there was a positive correlation between the duration of female care and the survival of larvae (Pearson, r = 0.39; p =0.00). During the intra-buccal and free larvae care, the resident males were significantly more aggressive towards male intruders (Kruskal-Wallis test, H = 36.08; p = 0.00) (Figure 3).

During free larval care, the males were significantly more aggressive towards the intruders of the same-sex (Kruskal-Wallis test, H = 18.27; p = 0.00). For the females there was no significant difference in relation to aggression towards either sex (Kruskal-Wallis test, H = 2.58; p = 0.10). However, there was no correlation between the pairs aggressiveness towards intruders and survival of free larvae (Pearson: for males p = 0.71; for females p = 0.62).

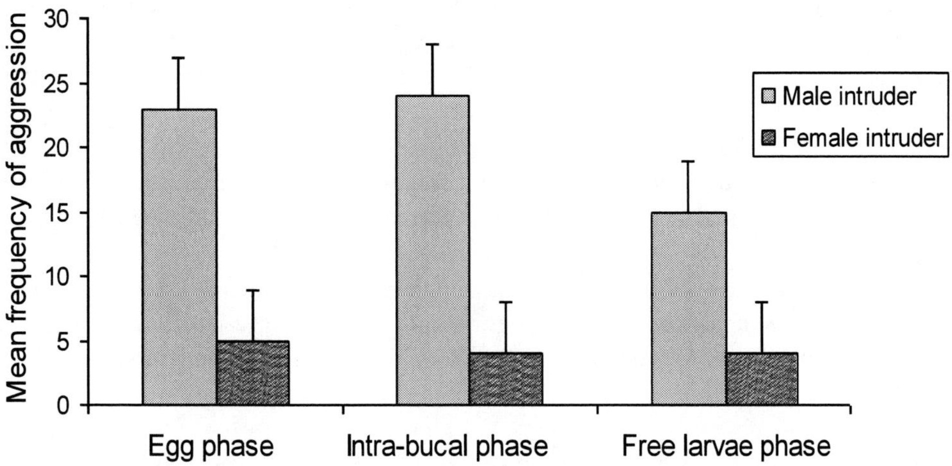

Figure 3. Mean frequency of aggression presented by resident males in the presence of intruder male and female during the phases of intra-bucal and free larvae care (Kruskal-Wallis test, H= 36.08; p<0.05).

Gonadal Maturation

The ovaries of hybrid red tilapia showed distinct stages of oocyte development which was consistent along the whole length of the ovary depending on the degree of ovarian maturation (Figure 4).

Stage I – Immature: During the growth stage, young germ cells compactly filled the ovaries, and occurrence of small sized germ cells were observed (phase I) along with oocytes of phase II (reserve stock). The ovogonia showed a well-defined spherical nucleus with clear nucleoplasm, thin filaments of chromatic material, with two or more nuclei. As the ovogonia grew, the initially clear cytoplasm acquired a basophilic nature, which became accentuated with the development of the cell. The immature oocytes had basophilic cytoplasm and were characterized by the total absence of yolk.

Stage II - Maturing: The ovaries of maturing females showed two distinct stages, the initial and final maturing. During the initial stage the oocytes became visible to the naked eye. As the oocytes developed, their colour gradually became reddish-pink, and at times even yellow. Final maturing was characterized by the presence of the standing stock of oocytes (phase II), as well as the oocytes with lipid droplets (phase III). This phase was characterized by the presence of big cells, voluminous nucleus and granular cytoplasm, with the appearance of cortical alveoli in the periphery. The cytoplasm showed presence of lipid droplets. The follicular layer showed an external surface with cells flatter than the theca. Final maturing stage was characterized by the bigger diameter of the oocytes due to the presence of lipid droplets and protein (phase IV) and some showed complete yolk formation (phase V).

Stage III- Mature: This stage showed oocytes in different stages of development (phases I to V) with an abundance of oocytes in phase V (yolk stage completed). The mature females showed the simultaneous occurrence of oocytes of different sizes in all developmental stages (Figure 4).

Stage IV – Partially Spent: The ovaries of partially spent females were flaccid with some haemorrhaging areas and empty space. Presence of post-ovulatory follicles, residual oocytes in the reabsorbing process of atresia, and immature and maturing oocytes characterized the partially spent stage.

Histological features of the testicular development in males of hybrid red tilapia showed immature, maturing, mature and partially spent stages. The germinal epithelium of the testes was composed of four main types germ cells: spermatogonia, spermatocytes, spermatids and spermatozoa.

Stage I – Immature: The seminiferous tubules showed the presence of primary and secondary spermatogonia. The primary spermatogonia were large and occurred in cysts all along the seminiferous tubule while the secondary spermatogonia were smaller in size with dark nucleus.

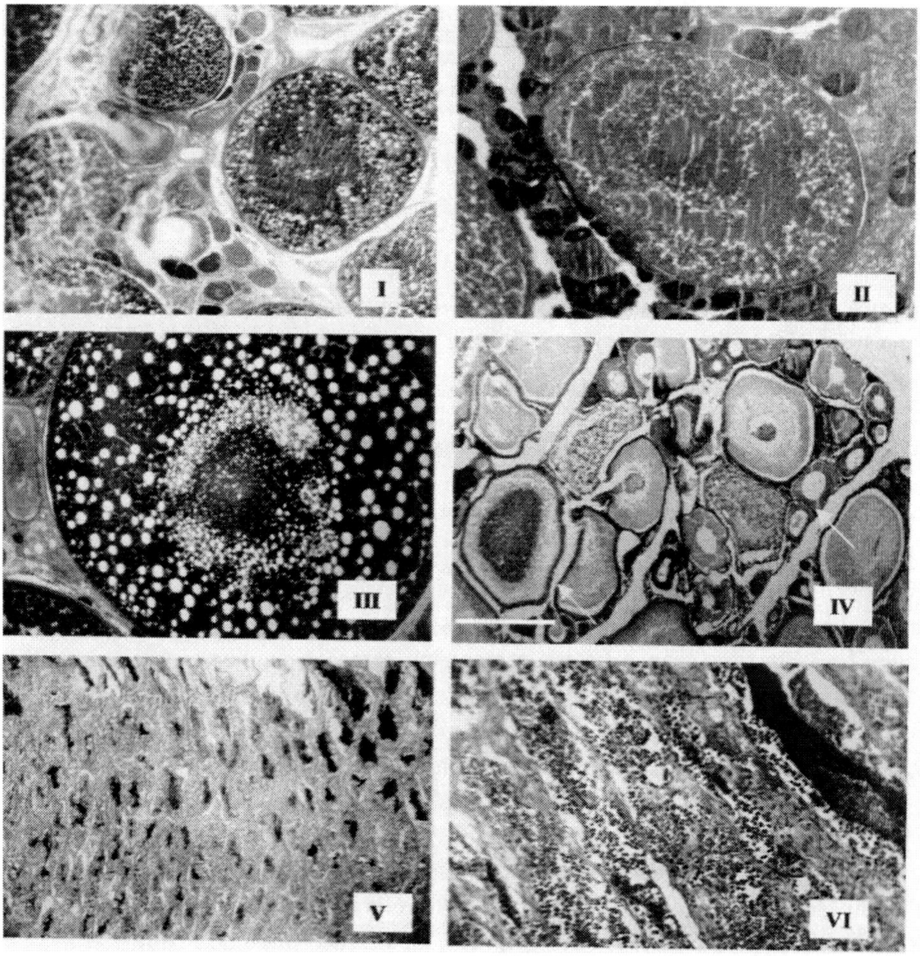

Figure 4. Histological sections showing gonad development in hybrid red tilapia. I to IV – ovaries (200x): I and II) oocytes in the maturation stage; III) mature oocyte showing complete vitologenesis; IV) partially spent stage, showing empty follicles (scale bar 200 μm), arrows pointing to immature and maturing oocytes. V and VI – testes (100x): V) immature with spermatogonia; VI) maturing stage with predominance of spermatocytes and spermatids.

Stage II - Maturing: The seminiferous tubules were bigger, with cysts of germ cells in advanced stages of maturation. Primary and secondary spermatocytes and spermatids were observed. During the advanced part of this stage, spermatozoa and lesser number of spermatogonia were observed.

Stage III- Mature: In mature testes, the lobules were large and the seminiferous duct was highly vascularized. The seminiferous tubules were packed with spermatozoa, some spermatids and spermatogonia.

Stage IV – Partially Spent: The testes were characterized with residues of germ cells, with the presence of spermatogonia in sufficient quantities and empty spaces.

Type of Spawning and Fecundity

Ovaries of mature females presented a mean batch fecundity of 1.213 mature oocytes. Mature ovaries had several batches of oocytes at different stages of development, indicating multiple spawning.

DISCUSSION

During reproduction, aggressive competition among males is common, in order to maximize the genetic contribution for future generations. In spite of differences in the movements that fish use when fighting, the end result in broad terms may be quite similar. Alternative strategies for resource acquisition, dominance hierarchies, territoriality and mate guarding are commonly encountered during the reproductive phase among fish (Huntingford and Chellappa, 2011). Fish often adapt the form and intensity of their agonistic responses to the context in which an encounter occurs. Thus dominance is one form of behaviour employed during conflict resolutions (Van Dorn et al, 2000). In fish, dominance is determined by individual attributes associated with body length, weight and level of aggressiveness (Whitherman and Côté, 2004). For exemple, during the reproductive season the large male cichlids in general are more aggressive and establish their dominance (Chellappa, 2000).

Dominance among hybrid red tilapia males was determined by the difference in body size and by aggressiveness. The aggression was directed from a large to a small individual, as was observed by Witheman and Côté (2004), studying the goby *Elacatinus prochilos.* According to these authors, dominance in this species was directly related to physical attributes. Large individuals were more aggressive and initiated most of the competitions. Dominance in territorial fish is generally identified by analysing agonistic profile, where the dominant attacks and the subordinate retreats (Ridley, 1995).

The gonadosomatic indices of hybrid red males showed a significant correlation with dominance level, with the dominant fish having higher indices than those of the subdominants and submissives. Gonad development may precede hierarchy establishment in some animals with higher GSI generally become dominant. Melamed et al (1999) observed that in the species *Oreochromis niloticus* the males with developed gonads have elevated plasma testosterone levels. This hormone is associated both to reproductive functions and increased aggressiveness in teleostei fish (Carlson et al., 2000). Thus, the higher GSI in the dominants, may be associated to the greater aggressiveness of these animals, and may affect their hierarchical relation in the group.

In hybrid red tilapia, nests excavated by the dominant males are circular in shape. The size of nests built by the males corresponded approximately to the total length of their bodies. This finding was also observed by Uglem and Rosenqvist (2002) in a study performed with *Symphodus melops*, where nest size was positively correlated with male body size.

In hybrid red tilapia, the females prefer large and aggressive males, those which excelled in courtship behaviour by removing sand and by swimming in circles. Gozlan et al. (2003), studying *Leucaspius delineatus,* found similar results. The females of this species preferred males which were more evident and more aggressive during courtship behaviour. According to these authors, selection was initially based on courtship behaviour. The aggressiveness displayed by the males was decisive in mate selection, for when they are more aggressive it represents a direct sign of physical condition and paternal competence.

This study shows that hybrid red tilapia males were aggressive during offspring care, with elevated aggressiveness displayed during intra-buccal egg and larval care. During free swimming larval stage the parents did not pay much attention with offspring care. Although the aggressiveness exhibited by the males reached high levels, the aggressiveness displayed by the pair was less than that of the intra-buccal phase.

This finding is likely related to the status of the pair during the intra-buccal phase, when the male must protect the offspring by himself, since the female is unable to react aggressively to predators. A similar result was found by Yamamoto et al. (1999) in the angelfish, *Pterophyllum scalare.*

Investment in offspring care is greater among the cichlids, where the offspring is protected by one or both parents from the egg stage until the larvae become independent (Keenleyside, 1991). Parental care includes guarding the eggs in the mouth, as in the *Xenotilapia flavipinnis* (Smith and Wootton, 1995), guarding and aerating of the eggs laid on the leaves of aquatic plants as in the *Pterophyllum scalare* (Cacho et al., 1999).

During all the parental care phases, the males proved to be more aggressive before same-sex intruders. Jaroensutasinee and Jaroensutasinee (2003), studying *Betta splendes,* reported higher male aggressiveness towards male intrusion than towards female intrusion. According to these authors, the males may be less aggressive towards opposite sex intruders, since the females may become potential partners. According to Huntingford (1976), the males exhibit high aggression levels in order to drive away the male intruder, a likely competitor.

There was no correlation between the aggressiveness of pairs towards intruders and survival of free larvae. During the free larvae phase the female was concerned with care for a period of three days, after which she lost interest in the offspring, which provoked a drop in the survival rate.

However, males were not involved with care in this phase, because they were possibly preparing for new matings in order to increase their reproductive success. The disinterest of the female for the offspring in this phase may be explained by the large energy expenditure during intra-buccal care, a phase during which she does not feed. According to Smith and Wootton (1995), intra-buccal care reduces reproductive success through increased spawning intervals.

The females may also reject the offspring when presented with new mating perspectives. Since this hybrid is a multiple spawning species, the females are ready to reproduce again within 27 days after the first spawn. Therefore, the interruption of parental care precedes their return to the reproductive process (Kubitza, 2000).

The general morphology and microscopic examination of histological sections of gonads of the hybrid red tilapia indicates that it is similar as observed for other cichlids, such as the angelfish, *Pterophyllum scalare* (Dias and Chellappa, 2003), *Cichla monoculus* (Chellappa *et al.,* 2003) and the red discus, *Symphysodon discus* (Chellappa et al., 2005).

Most species of cichlids spawn up to four times during the reproductive cycle (Keenleyside, 1991). Partial spawning extends over a prolonged period, thus reducing competition among larvae and the juveniles. The large reserve of yolk in oocytes allows larger and more developed larvae, which are defended from predators by their parents, thus ensuring a higher survival rate of offspring and greater reproductive success.

ACKNOWLEDGMENTS

The authors wish to thank the National Council for Scientific and Technological Development (CNPq) for the financial support awarded during the study period.

REFERENCES

Alvarenga, C.M,.D. and Volpato, G.L. (1995). Agonistic profile and metabolism in alevins of the Nile tilapia. *Physiology and Behavior*, 57(1):75-80.

Barber. I., Nairn, D. and Huntingford F. A. (2001). Nests as ornaments: revealing construction by male sticklebacks. *Behavioral Ecology*, 12 (4): 390-396.

Barlow, G.W. (1991). Mating systems among cichlid fishes. pp. 173-190. In: Keenleyside, M. H. A. (Ed.). *Cichlid Fishes-Behaviour, Ecology and Evolution*. London, Chapman and Hall. 378p.

Cacho, M.S.R. F., Chellappa, S. and Yamamoto, M. E. (2006). Reproductive success and female preference in the Amazonian cichlid angel fish, *Pterophyllum scalare* (Lichtenstein, 1823). *Neotropical Ichthyology*, 4 (1): 87 - 91.

Cacho, M.S.R. F., Yamamoto, M. E. and Chellappa, S. (1999). Comportamento reprodutivo do Acará bandeira, *Pterophyllum scalare* (Osteichthyes, Cichlidae). *Revista Brasileira de Zoologia*, 16: 653-664.

Cacho, M.S.R. F., Yamamoto, M. E. and Chellappa, S. (2007). Mating system of the Amazonian cichlid angel fish, *Pterophyllum scalare*. *Brazilian Journal of Biology*, 67 (1): 631 – 637, 2007.

Carlson, A. B., Hopkins, C.D. and Thomas, P. (2000). Androgen correlates of socially induced change in the electric organ discharge waveform of a mormyrid fish. *Hormonal Behavior*, 38: 177-186.

Chellappa, S. (2000). A Review on reproductive strategies and ecology of cichlid fishes of northeastern Brazil. *Revista Ecologia Aquática Tropical*, 10, 5-11.

Chellappa, S.; Câmara, M. R.; Verani, J. R. (2005). Ovarian development in the Amazonian red discus, *Symphysodon discus* Heckel (Osteichthyes: Cichlidae). *Brazilian Journal of Biology,* 65(4), 609-616.

Chellappa, S., Câmara, M. R., Chellappa, N. T., Beveridge, M. C. M. and Huntingford, F. A. (2003). Reproductive ecology of a neotropical cichlid fish, *Cichla monoculus* (Osteichthyes: Cichlidae). *Brazilian Journal of Biology*, 63:17-26.

Chellappa, S., Chellappa, N. T., Silva, E. A., Huntingford, F. A. and Beveridge, M. C. M. (1996). The diet of hybrid red tilapia, *Oreochromis niloticus* (Linnaeus) x *Oreochromis mossambicus* (Peters) reared in the freshwater ponds of Northeastern Brazil. *Aquaculture Research*. 27, 945-952.

Chellappa, S., Yamamoto, M.E. and Cacho, M.S.R.F. (1999a). Reproductive behaviour and ecology of two species of cichlid fishes. In: Vall A.L., Vall, V.M. (Eds.), *Biology of Tropical Fishes*. INPA, Manaus, p. 113-126.

Chellappa, S., Yamamoto, M.E., Cacho, M.S.R.F. and Huntingford, F.A. (1999b). Prior residence, body size and the dynamics of territorial disputes between male freshwater angelfish. *Journal of Fish Biology*, 55 (6), 1163-1170.

Dias, R. L. and Chellappa, S. (2003). Desenvolvimento gonadal do ciclídeo ornamental acará bandeira, *Pterophyllum scalare*. *XII Simpósio Brasileiro de Aqüicultura*. Vol. 2. (Eds. Urbinati, E.C. and Cyrino, J.E.P.) Jaboticabal, SP. p.135-148.

Fernandes, M.O. and Volpato, G.L. (1993). Heterogeneous growth in Nile tilapia: social stress and carbohydrate metabolism. *Physiology and Behavior*, 54: 319-323.

Gozlan, R.E., Flower, C.J. and Pinder, A.C. (2003). Reproductive success in male sunbleak, a recent invasive fish species in the U.K. *Journal of Fish Biology*, 63 (1):131-143.

Huntingford F.A. and Chellappa, S. (2011). Agressão. p. 191–207. In: M. E. Yamamoto and G. L. Volpato (Eds.), *Comportamento Animal*. Second edition, EDUFRN, Natal, Brazil. 342p.

Huntingford F. A. (1993). Behaviour, Ecology and Teleost Fishes. Pp. 1 – 5. In: Huntingford, F.A., Torricelli, P. (Eds.). *Behavioural Ecology of Fishes*. Harwood Academic Publishers, Chue. 322p.

Huntingford, F.A. (1976). A comparison of the reaction of sticklebacks in different reproductive conditions towards conspecifics and predators. *Animal Behaviour*. 24: 694-697.

Joaroensutasinee, M. and Joaroensutasinee, K. (2003). Type of intruder and reproductive phase influence male territorial defence in wild-caught Siamese fighting fish. *Behavioural Processes*, 64: 23-29.

Keenleyside, M. H. A. (1991). *Cichlid Fishes-Behaviour. Ecology and Evolution*. London, Chapman and Hall. 378p.

Kubitza, F. (2000). *Tilápia: Tecnologia e Planejamento na Produção Comercial*. Ed. USP, SP. 285p.

Kullander, S.O. (1998). A phylogeny and classification of the South American Cichlidae (Teleostei: Perciformes). p. 461-498. In L.R. Malabarba, R.E. Reis, R.P. Vari, Z.M. Lucena and C.A.S. Lucena (Eds.) *Phylogeny and classification of Neotropical fishes*. Porto Alegre, EDIPUCRS. 603 p.

Melamed, P., Gur, G., Rosenfeld, H., Elizur, A. and Yaron, Z. (1999). Possible interations between gonadothrophs and somatotrophs in the pituitary of tilapia: apparent roles for insulin-like growth factor I and estradiol. *Endocrinology*, 140 (3): 1183-1191.

Palmer, E.E., Sorensen, P.W. and Adelman, I.R. (1995). A histological study of seasonal ovarian development in freshwater drum in the Red Lakes, Minnesota. *Journal of Fish Biology*. 47, 199-210.

Ridley, M. (1995). *Animal Behaviour*. Blackwell Science Scientific Publication, USA. 288p.

Rogers, W. (1995). Female choice predicts the best father in a biparental fish, the Midas cichlid (*Cichlasoma citrinellum*). *Ethology*, 100: 230 – 241.

Smith, C. and Wootton, R. J. (1995). Experimental analysis of some factors affecting parental expenditure and investiment in *Cichlassoma nogrofasciatum* (Cichildae). *Enviromental Biology of Fishes*, 42: 289-302.

Uglem, I. and Rosenqvist, G. (2002). Nest bulding and mating in relation to male size in corkwing wrasse, *Symphodus melops*. *Enviromental Biology of Fishes*, 53 (1): 17-25.

Van Dorn, G. S., Hengeveld, G. M. and Weissing, F. J. (2000). The evolution of social dominance II: multi-Player Models. *Behaviour*, 140 (10): 1333-1358.

Vazzoler, A.E.A.M. (1996). *Biologia da Reprodução de Peixes Teleósteos: Teoria e Prática*. Paraná, Editora UEM, 169p.

Volpato, G.L. and Fernandes, M.O. (1994). Social control of growth in fish. *Brazilian Journal of Medical and Biological Reserch*, 27: 797-810.

West, G. (1990). Methods of assessing ovarian development in Fishes: a review. *Australian Journal of Marine and Freshwater Research*, 41, 199-222.

Whitheman, E. A. and Côté, I. M. (2004). Dominance hierarchies in group living cleaning gobies: causes and foraging consequences. *Animal Behavior*. 67: 239- 247.

Wootton, R.J., Evans, G. W. and Mills, L. A. (1978). Annual cycle in female three-spined sticklebacks (*Gasterosteus aculeatus*) from an upland and lowland population. *Journal of Fish Biology*. 12, 331-343.

Yamamoto, M.E., Chellappa, S., Cacho, M.S.R.F. and Huntingford, F. A. (1999). Mate guarding in an Amazonian cichlid, *Pterophyllum scalare*. *Journal of Fish Biology*, 55: 888-891.

SECTION III: ANATOMY AND HISTOLOGY OF THE DIGESTIVE TRACT AND FEEDING HABITS OF FISH SPECIES

In: Biology of Semiarid Tropical Fish
Editors: J. R. Gomes and S. Chellappa
ISBN: 978-1-63117-067-6

Chapter 6

ANATOMY AND HISTOLOGY OF THE DIGESTIVE TRACT AND FEEDING HABITS OF THE MARBLED SWAMP EEL, *SYNBRANCHUS MARMORATUS*

***Luciana Araújo Montenegro*[1], *Naisandra Bezerra da Silva*[2], *Wallace Silva do Nascimento*[3] and *Sathyabama Chellappa*[3*]**

[1]Departamento de Oceanografia e Limnologia, Centro de Biociências, Universidade Federal do Rio Grande do Norte, Praia Mãe Luíza, s/n, Natal, Rio Grande do Norte, Brazil

[2]Departamento de Morfologia, Centro de Biociências, Universidade Federal do Rio Grande do Norte, Lagoa Nova, Natal, Rio Grande do Norte, Brazil

[3]Postgraduate Program in Psychobiology, Centro de Biociências, Universidade Federal do Rio Grande do Norte, Lagoa Nova, Natal, Rio Grande do Norte, Brazil

ABSTRACT

The gastrointestinal tract of the marbled swamp eel, *Synbranchus marmoratus* has striking differences from the other fish, due to its aerial repiratory habits. This study investigated the food habits, anatomy and histology of the digestive tract of this species captured from a tropical reservoir in Northeastern Brazil. The stomach contents were analyzed using volumetric and frequency of occurrence methods besides the dietary importance index. *S. marmoratus* is a carnivore with preference to prawns. The anatomical and histological aspects of the digestive tract shows a ventro-terminal mouth, small teeth forming various series on maxilla, branchial arches without anatomically visible gill rakers, a short cylindrical thick walled oesophagus with mucas surface and internal parallel folds, stomach with cardiac, fundic and pyloric portions. The cardiac portion of the stomach is short and cylindrical formed of simple epithelial cylindrical mucus cells. The fundic portion is long with narrow walls, a big cavity and smaller folds which give rise to gastric glands. The pyloric portion has no glands and lacks primary and secondary mucus folds. The stomach and intestine are well vascularized to serve a role in

* E-mail: chellappa.sathyabama63@gmail.com.

respiration during estivation. The digestive tract of *S. marmoratus* shows adaptations to a carnivorous feeding habit and gastrointestinal respiration.

Keywords: *Synbranchus marmoratus*, anatomy, histology, carnivorous feeding habit, gastrointestinal respiration

INTRODUCTION

Information on feeding habits of fish in their natural habitat is essential to understand their biology and ecology. It helps to explain variations in growth, reproduction, ecological adaptations and migration besides the foraging behavior of fish (Murray et al., 1994; Little et al., 1998; Esteves et al., 2008, Ballesteros et al., 2009). Investigations of the food spectrum, through analysis of stomach contents of fish, coupled with anatomical and histological investigations of the digestive tract, have aided in interpreting the dynamics of feeding and habitats occupied by fish species (Hahn et al., 1997).

Plasticity in the diet of fish obeys predetermined limits imposed by the digestive tract, whose structures may reflect the food trends of the species (Fugi et al., 1996). Anatomical and histological studies of the digestive tract are useful for the characterization of the digestive organs with a better understanding of the physiology and feeding habits of the species (Park and Kim, 2001; Hernández et al., 2009). Studies on the description of the anatomical and histological structure of the digestive system of fish were carried out in South America by some authors (Grau et al., 1992; Morrison and Wright Jr., 1999; Silva et al. 2005; Canan et al., 2012).

The marbled swamp eel, *Synbranchus marmoratus* Bloch 1975 (Synbranchiformes: Synbranchidae), is a native fish of the Brazilian semiarid region. The Synbranchiformes fish have a long serpentine body lacking pelvic fins. A dominant characteristic of this species is hermaphroditism presenting sex reversal sequence, which is considered as a specialized capacity of this species (Lo Nostro and Guerrero, 1996). At the beginning of the dry season when the water level declines, *S. marmoratus* remains buried in the sediment (Lo Nostro, 2000). Fish living under semi arid climatic conditions are faced with several problems, and the most obvious of which is the lack of water. During the dry seasons, when the rivers and reservoirs dry up or have limited water sources, *S. marmoratus* is subjected to increasing environmental stress. The response of this species to dry conditions is remarkable. They bury themselves into the river or reservoir bed by constructing a mucus cocoon surrounding them. Metabolic processes are greatly reduced and the fish enter a state of aestivation, a form of extreme hibernation (Montenegro, 2009; Montenegro et al., 2011).

In some species of bimodal breathing fish, parts of the gastrointestinal tract, notably the stomach and/or portion of the small intestine, become modified to serve a role in respiration. The part of the digestive tract functioning as an air breathing organ becomes highly vascularized and this facilitates gas exchange between the blood and the swallowed air. Air is then voided either via the anus or via the mouth. Members of the families of armoured cat fishes (Callichthydae and Loricaridae) that inhabit freshwaters of South and Central America offer typical examples of gastrointestinal air breathers. *S. marmoratus* breathes air during terrestrial excursions and while dwelling in hypoxic conditions of water and utilizes its gills and adjacent buccopharyngeal epithelium as an air-breathing organ. This fish uses gills and skin for aquatic respiration in normoxic (air-saturated) water but when exposed to progressive aquatic hypoxia it becomes a metabolic oxygen conformer until facultative air breathing is initiated (Graham and Baird, 1984).

Considering the lack of scientific information relating to feeding habits, anatomy and histology of the digestive tract of *S. marmoratus*, this work aims to describe the feeding habits besides anatomical and histologicl characterization of the digestive tract of this species. This study contributes to the understanding of the food habits exhibited by the species during the rainy and dry seasons, and its adaption for aerial respiration using the digestive tract.

MATERIALS AND METHODS

Study Site and Fish Sampling

Fish sampling was carried out on a monthly basis from August, 2007 to July, 2008, in the Marechal Dutra Reservoir, RN, located in the Caatinga ecoregion of Northeastern Brazil (6° 26' 11''S and 36° 36' 17''W). The reservoir has an accumulation capacity of $40x10^6$ m^3 with maximum depth of 25m and mean depth of 8.5m, and is a source of potable water, besides being important for fisheries and agriculture. This reservoir was constructed on River Acauã in the hydrographic basin of River Piranhas-Assu, which is the main hydrographic basin in Rio Grande do Norte, covering an area of 44,000 km^2 (Chellappa et al., 2009).

Fish samples were captured with the help of fishermen, who used artisanal fish traps locally known as "covos" which were placed on the substrate close to the marginal vegetation. The fish traps were allowed to remain there stationary for 24 hours.

In order to characterize the dry and rainy seasons of the study area, rainfall data of the region during the study period was obtained from the Meteorological Department of EMPARN (Agriculture Research Company of Rio Grande do Norte), Natal, Brazil.

Measurements

Fish captured were numbered, measured, weighed and samples of fish were used for morphometric measurements to confirm the taxonomical identification of *S. marmoratus*. Measures of total length (Lt ± 1cm), total body mass (Wt ± 1 g) and stomach weight (Wg ± 0.1mg) were recorded. In this study a total number of 78 samples of the study species were used, with standard body length varying from 26.6 cm to 103.8 cm, with a mean of 66.5 (± 0.95) cm. The body weight varied from 19 g a 271.5 g with a mean of 43.1 (± 8.6) g.

Stomach Contents Analyses

Each fish was ventrally incised to expose the digestive tract which was removed from the coelomic cavity and fixed in 10% formaldehyde, and later conserved in 70% alcohol. The stomach contents were examined by the method of frequency of occurrence (Hynes, 1950; Hyslop, 1980). This corresponds to the percentage frequency of the stomachs where occurs the food item in relation to the number of stomachs with food and volume (Hyslop, 1980). The volume is expressed as percentage, considering the volume of food item in relation to the volume of all the food items present in the stomach.

Index of Alimentary Importance was applied on the values of frequency of occurrence and volume (Kawakami and Vazzoler, 1980), which provides the items that contribute most to the diet of the species. Alimentary Index IAi = Fo% x Vi% / Σ Fo% x Vi%, where: Fo = Frequency of occurrence (%) of a particular item; Vi = Volume (%) of a particular item.

The average stomach fullness index (IR) was analyzed (Hahn, 1991) which indicates a greater or lesser feeding activity due to availability or scarcity of food throughout the year. The average stomach fullness index IR = We X 100 / Wt, where: We = Stomach weight (g); Wt = Total body mass of fish (g).

Degree of Fullness of Stomachs was calculated (Berg, 1979; Gavilan-Leandro, 2003). The degrees of stomach fullness varied from 0 to 3, where: 0 = empty; 1 = up to 25% full; 2 = between 25% and 75% full; 3 = between 75% and 100% full.

For morphological descriptions, attention was given to the position and size of the mouth, type of dentition. The branchial arches and the gills rakers were examined.

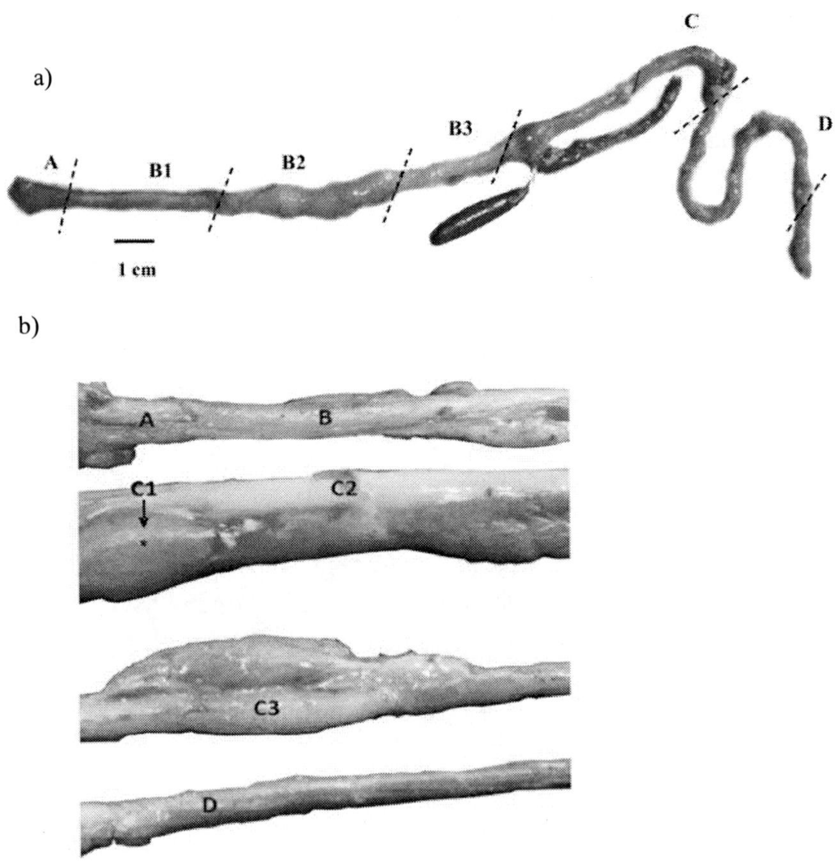

Figure 1 (a). Structures of the digestive tract of *Synbranchus marmoratus*: (A) esophagus; (B1) cardic stomach, (B2) fundic stomach, (B3) pyloric stomach, (C) anterior intestine (D) middle intestine. Figure 1 (b). Regions of the digestive tract of *Synbranchus marmoratus* submitted to microtomy: (A) esophagus; (B) cardic stomach (C1) * indicates the heart; (C2) fundic stomach; (C3) pyloric stomach; (D) anterior intestine.

Characterization of the Morphology of the Digestive Tract

The digestive tract was removed from the coelomic cavity from six samples of *S. marmoratus* for the characterization of the morphology of the digestive tract (Figure 1A and 1B). Tissue samples were fixed in Bouin's fluid. After 24 hours of fixation, the digestive tract was submitted to microtomy to obtain fragments of the digestive organs: esophagus, stomach (cardiac, fundic and pyloric regions), and the intestine (anterior, and middle portions). The fragments were submitted to routine histological techniques (dehydration, diaphanization and paraffin-embedding) and microsectioned at 5μm. Samples were stained with Haematoxylin-Eosin (Maia, 1979) and periodic acid Schiff (Behmer et al. 1976) following routine procedures. The sections were analyzed through transverse and longitudinal sections. Additionally, a sagittal section of the whole tract aided in the identification of transition regions. For anatomical analysis a stereoscope (PZO Labimex) was used. For the slide microphotography procedure an Olympus microscope CH30 coupled to an Olympus digital camera PM – 10 AK was used and the images were processed using Nikon ACT-1 software.

The following were also considered: the structures of the mucosa, the type of epithelium, and its component cell types, as well as possible occurrence of glands in the lamina propria and the presence or absence of muscular mucosa, the characterization of the submucosa, and distribution of granular cells or glands in the tunica and also the organization of the muscular layer of tunica, serosa or adventitia when present.

RESULTS

S. marmoratus has long, tubular body form with a ventro-terminal mouth. The thin lips are attached to the strong jaw, with small rounded teeth forming various series on maxilla. There are teeth even on the pharyngeal bones to function in the capture of food. There are four pairs of short and thick gill arches with the gill filaments on one side and the gill arches possessed no gill rakers (Figure 2).

Food and Feeding Habits

The analysis of stomach contents of *S. marmoratus* revealed that this fish fed exclusively on items of animal origin and therefore is regarded as a carnivorous species. The diet was composed of six items separated into distinct categories. The food items were composed of the crustacean *Macrobrachium amazonicum* Heller, 1862, besides fish, molluscs, nematodes and segments of insects. There were also organic material consisting of semi-digested fish scales, exoskeletons of crustaceans and large amount of debris. The Index of Alimentary Importance when applied to the total number of stomachs (n = 78) showed that the crustacean, represented by the family Palaemonidae, Decapoda, *M. amazonicum* was the most important item of the diet (Table 1); (Figure 3). The following matrials were observed in the stomach contents of *S. marmoratus*: organic material represented by semi-digested remains of scales, exoskeletons and debris, fish including vertebrae and fin rays, molluscs represented by gastropod of the generas *Pomacea* and *Biomphalaria*, and remnants of shells; segments of unidentified insects and unidentified nematodes (Table 1).

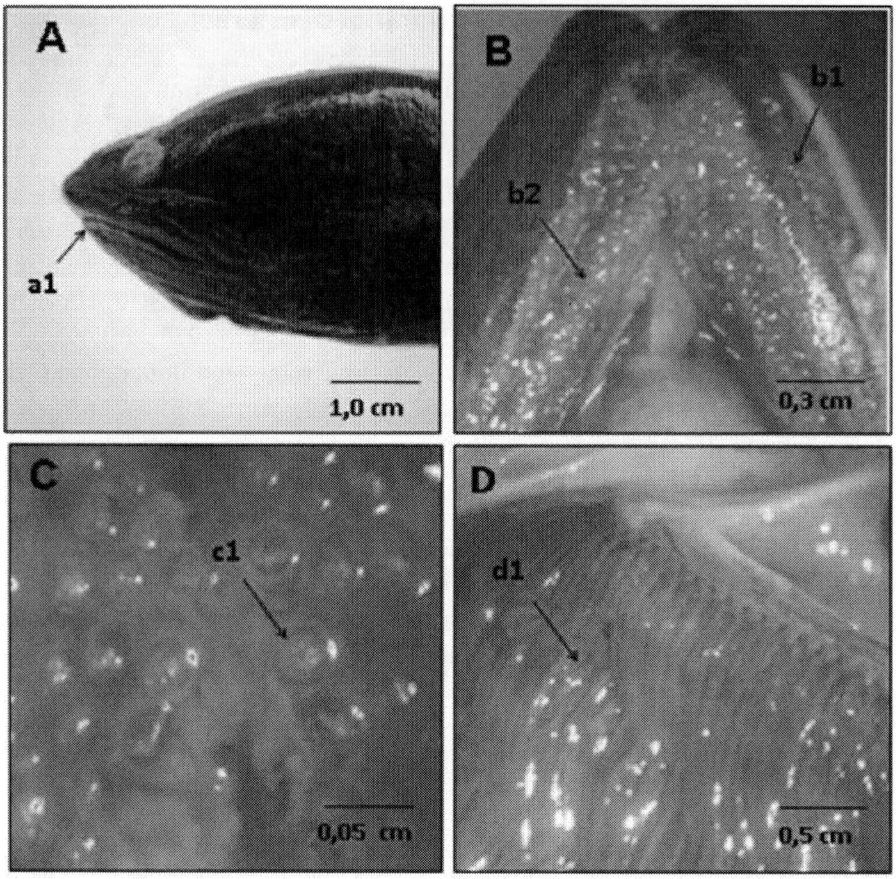

Figure 2. (A). Lateral view of the head of *Synbranchus marmoratus* a1) with ventro-terminal mouth; (B) a view of b1) the thin lips and b2) small teeth; (C) a closer view of the small teeth; (D). the first branchial arch showing d1) gill filaments without anatomically visible gill rackers.

Table 1. Food spectrum of *Synbranchus marmoratus* during (a) dry season (n=26) e (b) rainy season (n=52); Fo = Frequency of occurrence, Fv = Volumetric frequency (%) and IAi = Index of Alimentary importance, IAi % = Index of Alimentary importance (%)

a)

Food items	Fo	Fv %	IAi	IAi %
Nematode	8,0	3,9	0,0053	0,53
Crustacea (shrimps)	80,0	42,5	0,5862	58,62
Molluscs	36,0	11,4	0,070	7,0
Fish	36,0	16,0	0,0993	9,93
Remains of Insects	16,0	3,9	0,0107	1,07
Organic material (detritus)	60,0	22,0	0,2275	22,75

b)

Food items	Fo	Fv %	IAi	IAi %
Nematode	-	-	-	-
Crustacea (shrimps)	90,0	73,64	0,922672	92,2672
Molluscs	7,5	4,05	0,004228	0,4228
Fish	5,0	2,7	0,001879	0,1879
Remains of Insects	7,5	3,37	0,003518	0,3518
Organic material (detritus)	30,0	16,21	0,067701	6,7701

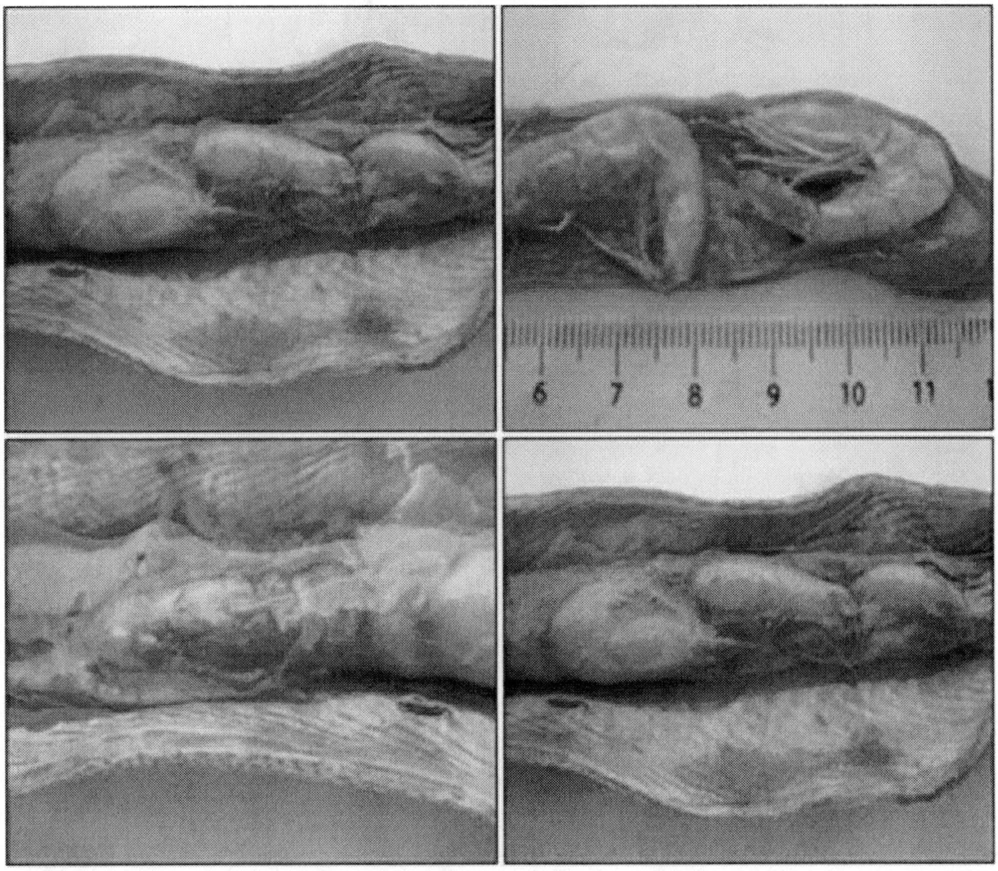

Figure 3. Stomach contents of *S. marmoratus* showing samples of the shrimp, *Macrobrachium amazonicum*.

Relationship between Feeding and Rainfall

During the study period, rainfall varied from 0 mm to 335.5 mm, with a mean annual rainfall of 118.01mm (± SD 13.10). The peak rainfall occurred in March (335.5 mm), and there were no rains during the months of August to December. The analysis of the mean stomach fullness index (RI) showed that the months from August to December, corresponding to the dry season, showed the highest values of stomach fullness, with a peak in October. There was a marked decrease in the RI in the months January to June. Feeding activity was higher during the dry season and a decrease was observed during the rainy season. Stomach fullness index verified that 16% of the stomachs were empty. During the dry season five stomachs were empty and four were full (Figure 4). During the rainy season eight stomachs were empty and none were full.

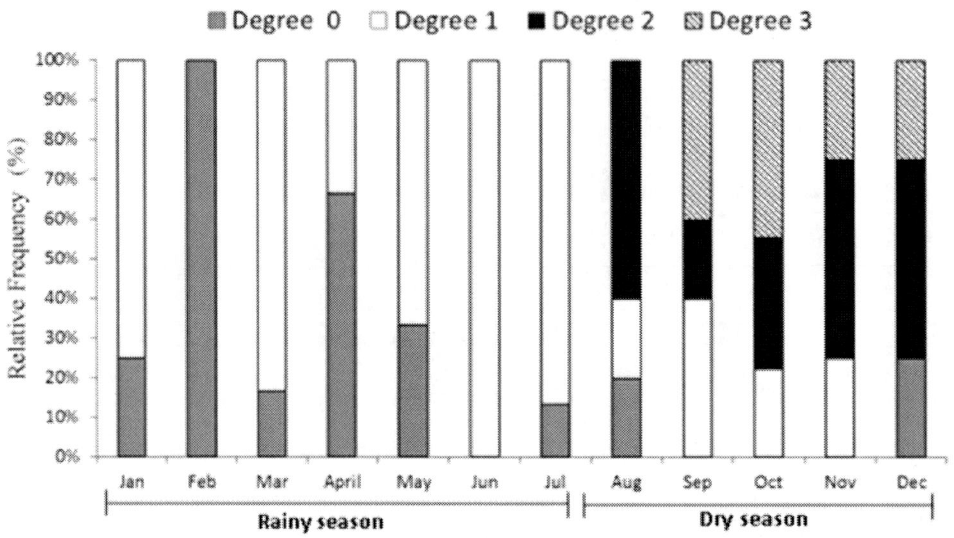

Figure 4. Stomach Fullness Index of *S. marmoratus* in relation to rainfall during Rainy season and Dry season.

Anatomy and Histology of the Gastrointestinal Tract

The esophagus *S. marmoratus* is short, cylindrical in shape and is of small diameter. The esophageal wall is thick, with the inner surface formed by mucosal lining containing parallel folds. The mucosa shows a pseudo stratified epithelium (Figure 5A) containing mucus-secreting cells, basal cells and apical lining cells. The presence of lamina propria was formed of dense connective tissue and glands were absent. The muscular layer consists of smooth and striated muscles. The esophagus is covered by a layer of adventitia and when it approaches the stomach appears the serosa. In the transition area between the two regions, epithelium gradually becomes cylindrical and the number of mucus-secreting cells increases (Figure 5B). The cuboidal epithelial cells lining the esophagus showed basal cells and mucous cells (Figure 5C). There are folds in the mucosa with bundles of striated muscle reaching the lamina propria (Figure 5D). The stomach is straight and long, with the cardiac, fundic and pyloric stomach regions. The following histological layers were observed in the stomach of *S. marmoratus:* mucosa, submucosa, muscular tunica and serosa. It features thick walls and reduced light due to the large number of folds that are present (Figure 6A).

The cardiac region, the first portion of the stomach, is short and cylindrical, formed by mucosa consisting of simple columnar epithelium with many mucous cells. This region presents gastric glands (Figure 6A) and prominently shows the appearance of a very vascular submucosa and serosa (Figure 6B and 6C). The muscular layer of cardiac region is formed by smooth muscle, dividing into two layers: the first is circular which is well developed in most regions, and more externally is a longitudinal layer.

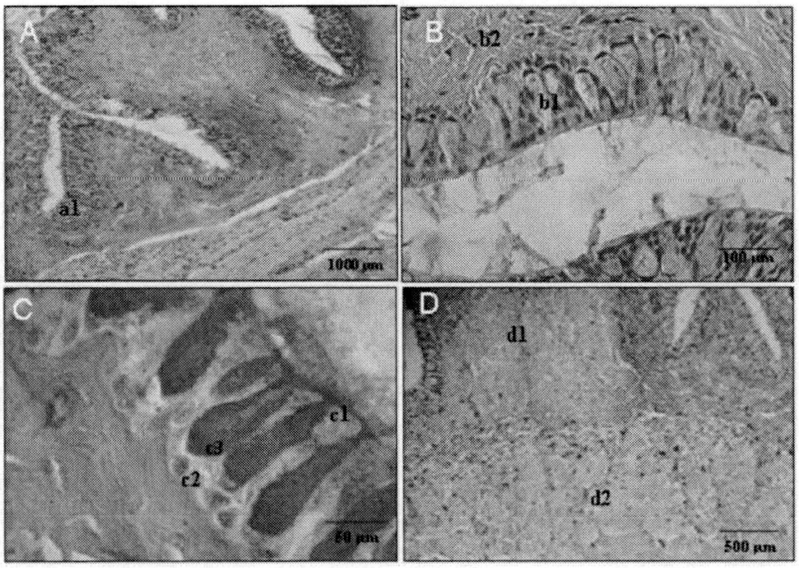

Figure 5. Esophagus of *Synbranchus marmoratus* showing: (A) a1) pseudo stratified epithelium (HE); (B) b1) mucosecretory pseudo stratified epithelium, b2) lamina propria (HE); (C) c1) cuboidal epithelial cells lining the esophagus, c2) basal cells, c3) goblet cells (PAS+); D d1) lamina propria of esophageal fold, d2) muscle layer containing striated muscular fibers (HE).

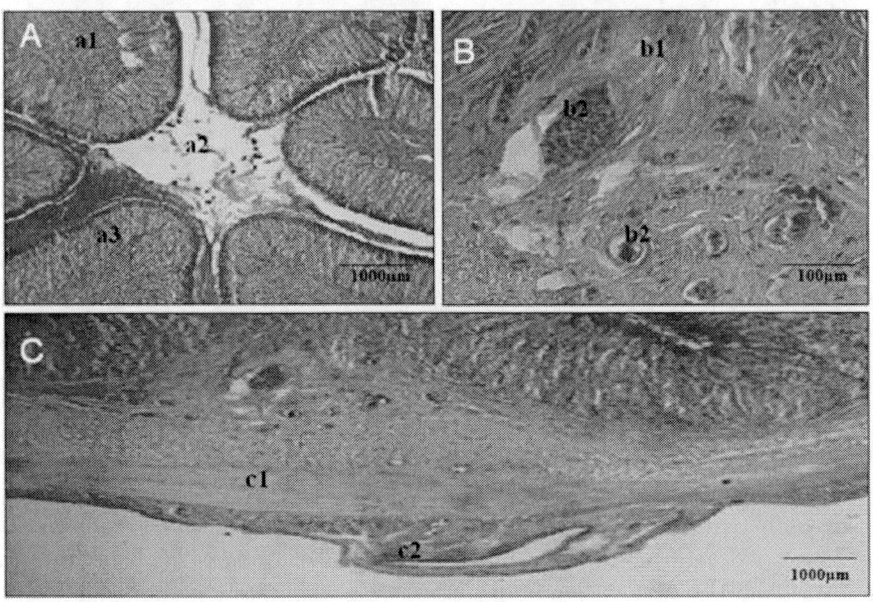

Figure 6. Cardiac stomach of *Synbranchus marmoratus*: (A) a1) submucosa, a2) light of the organ, a3) lamina propria with gastric glands (PAS); (B) b1) submucosa, b2) blood vessels; (C) c1) muscular layer, c2) vascularized serosa (PAS).

The stomach folds in fundic region consist of mucosa and submucosa (Figure 7 A). The submucosa did not show the presence of glands, and consists of connective tissue containing blood vessels. The fundic region is elongated to receive the food, and has thinner walls, greater light and folded smaller in proportion to the size. It is formed by simple columnar epithelium with mucous cells, it invaginates and gives origin to the gastric glands which secrete probably hydrochloric acid and pepsinogen (Figure 7B). The muscular layer of the fundic portion is divided into a longitudinal and a circular layer. The muscular layer is of smooth muscle, dividing into two layers, the first is circular, larger in most regions, and more externally a longitudinal layer and a highly vascualarized serosa (Figure 7C and 7D).

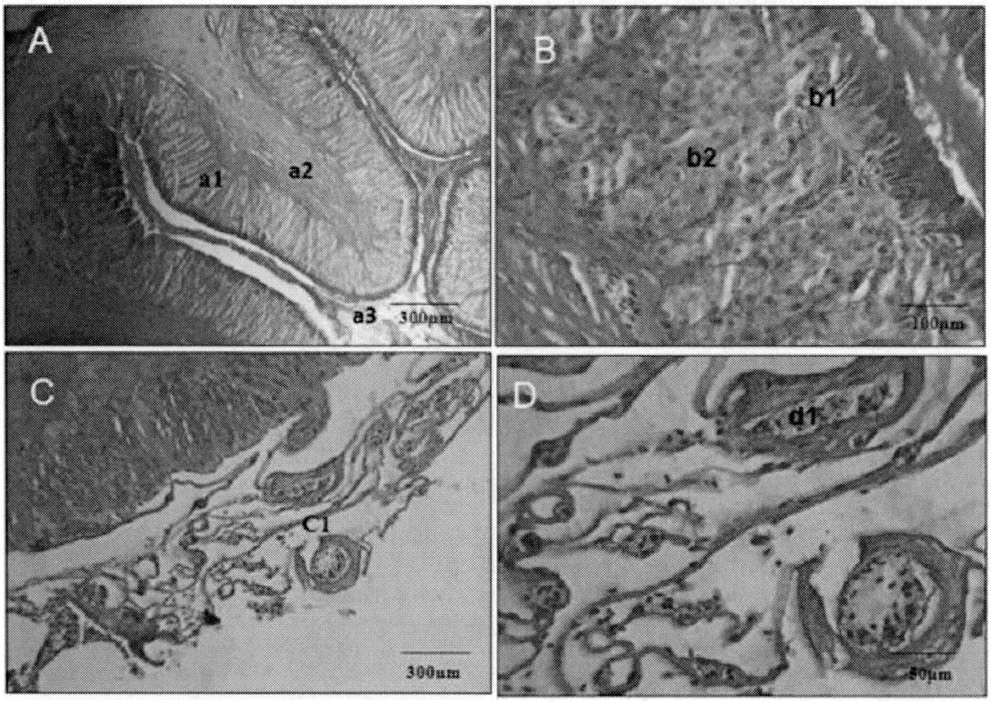

Figure 7. Fundic stomach of *Synbranchus marmoratus*: (A) folds of the stomach containing: a1) lamina propria with fundic glands, a2) submucosa without glands, a3) light containing mucous cells (PAS); (B) b1) simple cylindrical mucosecretory epithelium (PAS), b2) gastric glands; (C) c1) serous containing blood vessels (PAS); (D) d1) blood vessels with erythrocytes (PAS).

The pyloric mucosa is composed of simple columnar epithelium secreting mucus (Figure 8A), followed by lamina propria containing blood vessels only. The pyloric portion is without glands, in the mucosa there was the presence of main folds with small secondary folds (Figure 8B).

Throughout the pyloric portion, there are several small folds in the longitudinal direction of varying sizes. Along the primary folds, there are secondary invaginations giving an appearance of dentition and forming the coating epithelium with an ample surface for secretion. The circular muscle layer has a diameter greater than the other muscle layers of the stomach (Figure 8C).

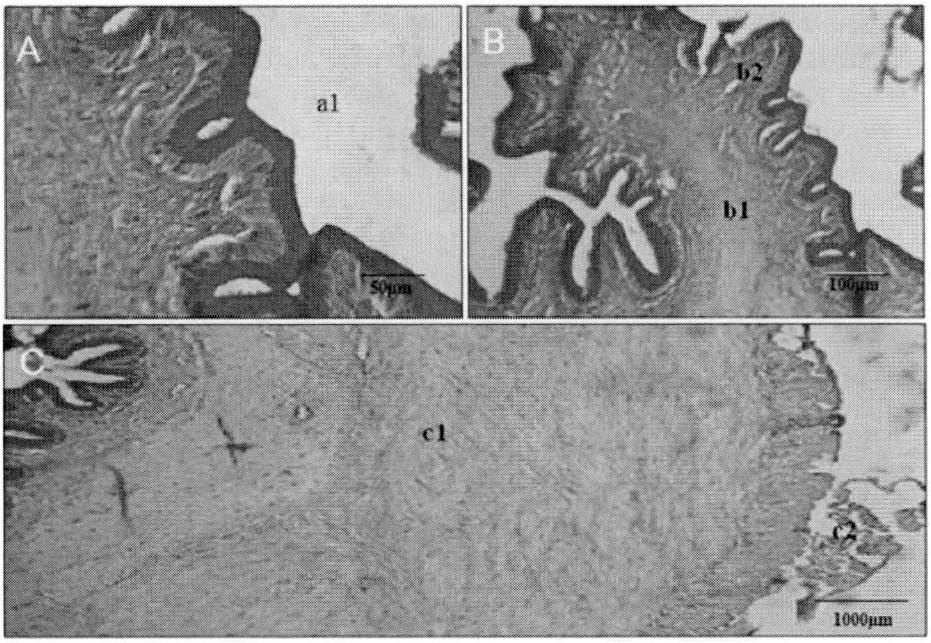

Figure 8. Pyloric stomach of *Synbranchus marmoratus*: (A) a1) light of the organ, a2) simple cylindrical mucosecretory epithelium (PAS); (B) b1) mucosal folds, b2) lamina propria without glands (PAS); (C) c1) muscular circular layer, c2) submucosa.

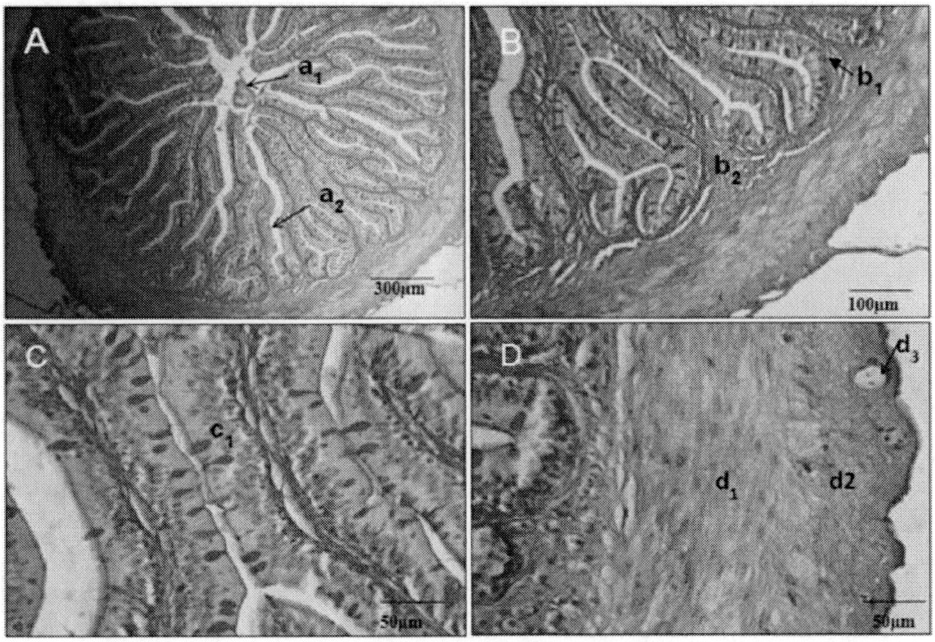

Figure 9. Intestine of *Synbranchus marmoratus*: (A) a1) intestinal light, a2) intestinal folds (PAS); (B) b1) simple cylindrical epithelium with goblet cells b2) lamina propria without glands (PAS); (C) c1) mucous secreting cylindrical cells (PAS); D d1) circular muscle layer, d2) longitudinal muscle layer d3) serous containing blood vessels (PAS).

There were variations in the presence of glands according to the region of the stomach. The cardiac region had small proportion of glands. In the fundic region a large numbers of gastric glands were present as a result of invaginations of the mucosal folds, with simple cylindrical epithelium and mucous cells. These persist in great quantities in the fundic region, and there was a decrease in the number until they completely disappear in the pyloric region. The submucosa is densely vascularized in the three regions. The stomach is externally covered by a serosa, where there are blood vessels and small nerves. The histological characteristics of the stomach of *S. marmoratus,* suggest that the stomach of this species is specialized in chemical digestion, like in carnivores besides facilitating in respiration.

Intestinal wall was composed of three layers: the mucosa (epithelial layer), lamina propria, muscularis and serosa. The histological analysis of the intestine shows the presence of intestinal villi consisting of simple columnar epithelium, which together with the lamina propria originate long and branched villi, lined by mucus-secreting cells (Figure 9C) and absorptive cells. The intestinal mucosa does not have glands and a submucosa was not observed (Figure 9A and B). Later on appears the muscularis consisting of tunica circular and longitudinal layers. The last layer is the serosa which contains a rich capillary network (Figure 9D).

Discussion

Fish show a wide adaptive radiation in their feeding habits and occupy every possible trophic role, from herbivorous to carnivorous levels (Keenleyside, 1979). They feed on the various resources found at different depths of water and their feeding habits along with the digestive tract descriptions provide useful ecological and biological information. In tropical aquatic ecosystems, most of the fish species show considerable plasticity in their diets despite their anatomical specializations for feeding (Lowe-McConnell, 1987, Mazzoni and Resende, 2003; Shibatta and Bennemann 2003). *S. marmoratus* has a ventro-terminal mouth indicating benthic food habits. The teeth are with rounded non-serrated edges which are useful for capturing and holding the prey. It is generally assumed that gill rakers function as a sieve to extract food organisms from the water and a relation exists between gill raker morphology and trophic status of the fish. In this study it was observed that the gill rakers were absent in *S. marmoratus*.

The diversity of the food items in the stomach of *S. marmoratus* permits to infer that this species is carnivorous with characteristics of an opportunistic species. Plasticity in the diet of fish, obeys a certain limit enforced by the macroscopic and microscopic structures of the digestive tract (Nikolsky, 1963; Watson and Balon, 1984; Weatherley and Gil, 1987; MacNeill and Brandt 1990; Cassemiro et al., 2003; Cinar and Enol, 2006).

The rainfall regime, food supply and population density may be considered as determinants of feeding activity, since the rains extend biotopes, thus favoring the feeding activity (Welcomme, 1995; Chellappa et al., 1996; Gil and Hart 1999). The analysis of stomach contents of *S. marmoratus*, during the periods of rain and drought, showed that the highest values for the stomach fullness index occurred during the dry season. This suggests that there was higher feeding activity during the dry season and *S. marmoratus* feeds lesser during the rainy months. This is possibly related to the reproductive period of this species, since most tropical fish species spawn during the rainy season. Low feeding activity corresponds with the gonadal development activities (Chellappa et al., 2009). Seasonal variation of feeding in fish may reflect the behaviour of the prey or the result of competition for the same food (Zaret and Rand, 1972; Guillen and Hart, 1980; Resende, 2000; Melo et al., 2004). The high frequency of occurrence of prawns in the stomach contents of *S. marmoratus* during the dry months may be related to the abundance and behaviour of the prey. Probably the capture of prawns is facilitated during the dry season, due to their migration to the margins of the reservoir during this season. Prey selection also depends on the mechanical ability of the fish to manipulate the prey successfully.

Generally in the anterior part of the esophagus the skeletal striated fibers predominate and the posterior part is composed of smooth muscle fibers. However, in *S. marmoratus* the esophageal muscular layer consists of smooth and striated muscles over its full length and the mucosa shows a pseudo stratified epithelium containing mucous-secreting cells.

In fish, the stomach is an organ with varied shape and structure, depending on their food habits and in some species it may be absent. The stomach in most fish is a dilation of the digestive tube, where food is retained for the required time for acid digestion. The inner mucosa of fish stomach forms winding longitudinal grooves that disappear when the stomach expands with the entry of food. The cardiac and fundic regions are glandular and the pyloric region is non glandular. The development of the fundic region in *S. marmoratus*, could be related to its ability to attack and consume large quantities of prey in a short time.

The histological characteristics of the stomach of *S. marmoratus* suggest that this species is specialized in chemical digestion, being considered as a carnivorous species. In the Amazon lake Chameleon, the food spectrum of *S. marmoratus* consisted primarily of insects, mites, crustaceans and fish (Soares et al., 1986).

Among vertebrates, the teleost fish present the most simplified intestinal histology. The intestine of fish usually does not have anatomical features that permit dividing it into well defined segments (Al-Hussaini, 1949; Khanna, 1964; Fugi et al., 1996; Albrecht et al., 2005). Thus, to distinguish between the different regions of the intestine of *S. marmoratus* further histophysiological and histochemical studies will be required, which may allow differentiation, between the areas of fat absorption (duodenum), absorption of proteins (large intestine) and absorption of water and ions (rectum).

S. marmoratus shows vascularized stomach and intestine which suggests their important role in respiration. This may be very useful especially during the aestivation period of the fish. Fish that breathe air can be separated into two groups, amphibious and aquatic air breathers. Amphibious air breathers rely on this respiratory mode during their frequent and routine terrestrial excursions, whereas aquatic air breathers, which for the most part remain confined to water, periodically surface and gulp air to obtain supplemental oxygen (Graham, 1976). In an earlier study it was registered that *S. marmoratus* breathes air during terrestrial excursions and while dwelling in hypoxic conditions of water and utilizes its gills and adjacent buccopharyngeal epithelium as an air-breathing organ (Graham and Baird, 1984). In the present study, it has become evident that due to their vascularized nature the stomach and intestine of *S. marmoratus*, both play an important role in respiration.

Anatomical and histological aspects of the digestive tract of *S. marmoratus* show that this species is adapted to carnivorous feeding habits, since its digestive tract has structures similar to other carnivorous species. The vascularized stomach and intestine possibly plays an important role in respiration.

Acknowledgments

This study was supported by a grant to the first author by the Post-Graduate Federal Agency CAPES/MEC, Brazil. Thanks are due to the National Council for Scientific and Technological Development of the Ministry of Science and Technology of Brazil (CNPq/ MCT) for the Research grants offered to the third and fourth authors.

References

Albrecht, M. P., Ferreira, M. F. N. and Caramaschi, E. P. 2005. Anatomical features and histology of the digestive tract of two related Neotropical omnivorous fishes (Characiformes; Anostomidae). *Journal of Fish Biology*, 58: 419-430.

Al-Hussaini, A. H. 1949. On the functional morphology of the alimentary tract of some fish in relation to differences in their feeding habits. II–Cytology and Physiology. *Quarterly Journal of Microscopical Science*, 90: 323-354.

Ballesteros, T. M., Torres-Mejia, M. and Ramírez-Pinilla, M. P. 2009. How does diet influence the reproductive seasonality of tropical freshwater fish? A case study of a characin in a tropical mountain river. *Neotropical Ichthyology*, 7: 693-700.

Behmer, A. O., Tolosa, E. M. C. and Neto, A. G. F. 1976. *Manual de técnicas para histologia normal e patológica*. São Paulo: Editora da Universidade de São Paulo.

Berg, J. 1979. Discussion of methods of investigating the food of fishes, with reference to a preliminary study of the prey of *Gobiusculus flavescens* (Gobiidae). *Marine Biology*, 50: 263-273.

Canan, B., Nascimento, W. S., Silva, N. B. and Chellappa, S. 2012. Morphohistology of the digestive tract of the damsel fish, *Stegastes fuscus* (Osteichthyes: Pomacentridae). *The Scientific World Journal*, Article ID 787316: 1-9.

Cassemiro, F. A. S., Hahn, N. S. and Rangel, T. F. L. V. B. 2003. Diet and trophic ecomorphology of the silverside, *Odontesthes bonariensis*, of the Salto Caxias reservoir, rio Iguaçu, Paraná, Brazil. *Neotropical Ichthyology*, 1: 127-131.

Chellappa, S., Chellappa, N. T., Silva, E. A., Huntingford, F. A. and Beveridge, M. C. M. 1996. The diet of hybrid red tilapia, *Oreochromis niloticus* (Linnaeus) x *Oreochromis mossambicus* (Peters) reared in the freshwater ponds of Northeastern Brazil. *Aquaculture Research*, 27: 945-952.

Chellappa, S., Bueno, R. M. X., Chellappa, T., Chellappa, N. T. and Val, V. M. F. A. 2009. Reproductive seasonality of the fish fauna and limnoecology of semi-arid Brazilian reservoirs. *Limnologica*, 39: 325- 329.

Cinar, K. and Enol, N. S. 2006. Histological and Histochemical Characterization of the Mucosa of the Digestive Tract in Flower Fish (*Pseudophoxınus antalyae*). *Anatomia, Histologia, Embryologia,* 35: 147-151.

Esteves, K. E., Lobo, A. V. P. and Faria, M. D. R. 2008. Trophic structure of a fish community along environmental gradientes of a subtropical river (Paraitinga River, Upper Tietê River Basin, Brasil). *Hydrobiologia*, 598: 373-387.

Fugi, R., Hahn, N. S. and Agostinho, A. A. 1996. Feeding styles of five species of bottom-feeding fishes of the High Paraná River. *Environmental Biology of Fishes*, 46: 297- 307.

Gavilan-Leandro, S. A. C. 2003. Variação temporal da freqüência de captura e do comportamento alimentar de *Astyanax bimaculatus* Linnaeus, 1758 (Characidae, Tetragonopterinae) do Rio Apodi-Mossoró, Rio Grande do Norte. Doctoral Thesis, Universidade Federal do Rio Grande do Norte, Natal. p.193.

Graham, J. B. 1976. Respiratory adaptations of marine air-breathing fishes. In: Hughes, G. M., ed. *Respiration of Amphibious Vertebrates*, London: Academic Press.

Graham, J. B. and Baird, T. A. 1984. The transition to air breathing in fishes III. Effects of body size and aquatic hypoxia on the aerial gas exchange of the swamp eel *Synbranchus marmoratus*. *Journal of Experimental Biology*, 108: 357-375.

Grau, A., Crespo, S., Sarasguete, N. C. and Gonzales d' Canales, N. L. 1992. The digestive tract of the Amberjack, *Seriola dumerili*, Risso: A light and scanning electron microscope study. *Journal of Fish Biology*, 41: 387-390.

Gil, A. B. and Hart, P. J. B. 1999. Dynamic changes in prey choice by stickleback during simultaneous encounter with large prey. *Journal of Fish Biology*, 55: 1317-1327.

Guillen, A. L. and Hart, T. 1980. Feeding interrelationships between the sand shiner and the striped shiner. *Ohio Journal of Science*, 80: 71-76.

Hahn, N. S. 1991. Alimentação e dinâmica da nutrição da curvina *Plagioscion squamosíssimus* (Heckel. 1840) (Piscas, Perciformes) e aspectos da estrutura trofica da ictiofauna acompanhante no rio Parana. Rio Claro: UNESP. 1991. Doctoral Thesis, Universidade Estadual Paulista, p. 287.

Hahn, N. S., Agostinho, A. A. and Goitein, R. 1997. Feeding ecology of curvina *Plagioscion squamosissimus* (Heckel, 1840) (Osteichthyes, Perciformes) in the Itaipu Reservoir and Porto Rico floodplain. *Acta Limnologica Brasiliensia*, 9: 11-22.

Hernández, D. R., Gianeselli, P. M. and Domitrovic, H. A. 2009. Morphology, Histology and Histochemistry of the Digestive System of South American Catfish (*Rhamdia quelen*) *International Journal of Morphology*, 27: 105-111.

Hynes, H. B. N. 1950. The food of freshwater stickle back (*Gasterosteus aculeatus* and *Pungosteus pungitius*) with a review of methods used in studies of the food fishes. *Journal of Animal Ecology*, 19: 35-38.

Hyslop, E. J. 1980. Stomach contents analysis - a review of methods and their application. *Journal of Fish Biology*, 17: 411-429.

Kawakami, E. and Vazzoler, G. 1980. Método gráfico e estimativa de índice alimentar aplicado ao estudo de alimentação de peixes. *Boletim do Instituto Oceanográfico*, 29: 205-207.

Keenleyside, M. H. A. 1979. *Diversity and adaptation in fish behaviour*. Springer Verlag, Berlim, 208p.

Khanna, S. S. 1964. The alimentary canal in some teleostean fishes. *Journal of the Zoological Society of India*, 13: 206-219.

Little, A. S., Tonn, W. M., Tallman, R. F. and Reist, J. D. 1998. Seasonal variation in diet and trophic relationships within the fish communities of the lower Slaver River, Northwest Territories, Canada. *Environmental Biology of Fishes*, 53: 429 - 445.

Lo Nostro, F. L. 2000. Espermatogénesis, ciclo anual e inducción hormonal de la espermiación en el pez protogínico diándrico , *Synbranchus marmoratus*, Bloch, 1795 (Teleostei, Synbranchidae). Doctoral Thesis. Universidad de Buenos Aires, Argentina, p. 170.

Lo Nostro, F. L. and G. A. Guerrero, 1996. Presence of primary and secondary males in a population of the protogynous fish *Synbranchus marmoratus*. *Journal of Fish Biology*, 49: 788 - 800.

Lowe-Mcconnell, R. H. 1987. *Ecological studies in tropical fish communities*. Cambridge, Cambridge University Press. 382p.

Maia, V. 1979. *Técnica Histológica*. 2. ed. São Paulo: Atheneu. 70-136.

Murray, H. M., Wright, G. M. and Goff, G. P. 1994. A comparative histological and histochemical study of the stomach from three species of Pleuronectid, the Atlantic halibut, *Hippoglossus hippoglossus*, the yellowtail flounder, *Pleuronectes ferruginea*, and the winter flounder, *Pleuronectes americanus*. *Canadian Journal of Zoology*, 72: 1199 - 1210.

Morrison, C. M. and Wright Jr, J. R. 1999. A study of the histology of the digestive tract of the Nile tilapia. *Journal of Fish Biology*, 54: 597- 606.

MacNeill, D. B. and Brandt, S. B. 1990. Ontogenetic shifts in gill-raker morphology and predicted prey capture efficiency of the alewife *Alosa Pseudoharengus*. *Copeia*, Carbondale. 1: 164-171.

Mazzoni, R. and Rezende, C. F. 2003. Seasonal diet shift in a tetragonopterinae (Osteichthyes, Characidae) from the Ubatiba River, RJ, Brazil. *Brazilian Journal of Biology*, 63: 69 - 74.

Melo, C. E., Machado, F. A. and Pinto-Silva, V. 2004. Feeding habits of fish from a stream in the savanna of Central Brazil, Araguaia Basin. *Neotropical Ichthyology*, 2: 37-44.

Montenegro, L. A. 2009. Biologia alimentar e morfohistologia do tubo digestório do mussum, *Synbranchus marmoratus*, Bloch (Osteichthyes: Synbranchidae) no açude Marechal Dutra "Gargalheiras", localizado no semiárido brasileiro. Masters Dissertation. Universidade Federal do Rio Grande do Norte, UFRN, Natal, p.84.

Montenegro, L. A., Damasceno, D. N. F., Almeida, R. G. and Chellappa, S. 2011. Biologia alimentar do mussum, *Synbranchus marmoratus* (Bloch, 1795) (Osteichthyes: Synbranchidae) no açude Marechal Dutra localizado no semiárido brasileiro. *Biota Amazônia*, 1: 52 - 60.

Nikolsky, G. V. 1963. *The ecology of fishes*. London: Academic Press.

Park, J. Y. and Kim, I. S. 2001. Histology and mucin histochemistry of the gastrointestinal tract of mud loach, in relation to respiration. *Journal of Fish Biology*, 58: 861-872.

Silva, N. B., Gurgel, H. C. B. and Santana, M. D. 2005. Histologia do sistema digestório de sagüiru, *Steindachnerina notonota* (Miranda Ribeiro, 1937) (Pisces, Curimatidae), do Rio Ceará Mirim, Rio Grande do Norte, Brasil. *Boletim do Instituto de Pesca*. 31: 1-8.

Resende, E. K. 2000. Trophic structure of fish assemblages in the lower Miranda river, Pantanal, Mato Grosso do Sul State, Brazil. *Revista Brasileira de Biologia*, 60: 389-403.

Shibatta, O. A. and Bennemann, S. T. 2003. Plasticidade alimentar em *Rivulus pictus* Costa (Osteichthtyes, Cypridontiformes, Rivulidae) de uma pequena lagoa em Brasília, Distrito Federal, Brasil. *Revista Brasileira de Zoologia,* 4: 615-618.

Soares, M. G. M., Almeida, R. G. and Junk, W. J. 1986. The trophic status of the fish fauna in Lago Camaleão, a macrophyte dominated floodplain lake in the middle Amazon. *Amazoniana*, 9: 511-526.

Watson, D. J. and Balon, E. K. 1984. Ecomorphological analyses of fish taxocenes in rainforest of northern Borneo. *Jounal of Fish Biology*, 25: 371-384.

Weatherley, A. H. and Gill, H. S. 1987. *Feeding relations correlated functional Morphology, growth and size*. London: Academic Press.

Welcomme, R. L. 1995. *Fisheries ecology of flood plain Rivers*. London: Longman, 1973.

Zaret, T. M. and Rand, A. S. 1972. Competition in tropical stream fishes: Support for the competitive exclusion principle. *Ecology*, 52: 336-342.

In: Biology of Semiarid Tropical Fish
Editors: J. R. Gomes and S. Chellappa

ISBN: 978-1-63117-067-6

Chapter 7

ANATOMY AND HISTOLOGY OF THE DIGESTIVE TRACT AND FEEDING HABITS OF *HOPLIAS MALABARICUS* (BLOCH, 1794) (OSTEICHTHYES: ERYTHRINIDAE)

Emilly Kataline Rodrigues Pessoa*[1], *Naisandra Bezerra da Silva*[2], *Arrilton Araújo de Souza*[1] *and Sathyabama Chellappa*[1]*

[1]Programa de Pós-Graduação em Psicobiologia, Centro de Biociências, Universidade Federal do Rio Grande do Norte (UFRN), Lagoa Nova, Natal, Rio Grande do Norte, Brazil

[2]Departamento de Morfologia, Centro de Biociências, Universidade Federal do Rio Grande do Norte (UFRN), Lagoa Nova, Natal, Rio Grande do Norte, Brazil

ABSTRACT

Hoplias malabaricus is a Neotropical fish with ecological and commercial importance, and has a wide geographical distribution, occurring in almost all the hydrographic basins of South America. This study investigated the anatomy and histology of the digestive tract besides the feeding habits of this species captured from a tropical reservoir in Northeastern Brazil. The stomach contents were analyzed using the frequency of occurrence method besides the dietary importance index. The digestive tract of *H. malabaricus* shows a dorso-terminal mouth, sharp strong teeth, the first branchial arch with anatomically visible gill rakers, a spacious buccopharyngeal cavity, a dorsally flattened short tubular esophagus with great elasticity, stomach with cardiac, fundic and pyloric regions, and a short intestine which opens out through the anus. The cardiac region of the stomach is short and thick, formed of three to seven primary folds, high and rounded along the lamina propria. The fundic region of the stomach has a big cavity and smaller primary folds which give rise to gastric glands. The pyloric region presents numerous parallel and very narrow longitudinal folds. The intestine is short, with an

* E-mail of the corresponding author: chellappa.sathyabama63@gmail.com.

intestinal coefficient of 0.72 ± 0.09. The dietary importance index indicates that *H. malabaricus* feeds preferentially on fish (72.8%) and prawns (27.2%).

Keywords: *Hoplias malabaricus*, anatomy, histology, carnivorous feeding habit

Introduction

Anatomical and histological characterization of the digestive tract coupled with analysis of stomach contents, have aided in interpreting the dynamics of feeding, habitat occupation, nutritional status and adaptations of the digestive tract in relation to the feeding habits of fish (Wootton, 1999; Novakowski et al., 2004). The freshwater fish fauna of the Neotropical region is characterized by high species richness which contributes to almost 31% of fish diversity of the planet (Reis et al., 2003).

Though morphology and histology of the fish digestive tract related to diet has been an interesting topic of investigation, such information pertaining to the Neotropical freshwater fish is still incipient (Albrecht et al., 2001; Canan et al., 2011; 2012; Silva et al., 2012; Montenegro et al., 2012).

The structure of the digestive tract of fish is related to their diet, which can be demonstrated when comparisons of digestive tract morphology are made between fish species of different trophic levels, especially carnivores and herbivores. The latter have long intestines, due to the difficulty of digesting plant carbohydrates and the amount of non-digestible matter, while the main characteristics of the digestive tract of carnivorous fish are big stomachs and relatively short intestines (Kapoor et al., 1975; Abdulhadi, 2005; Peretti and Adrian, 2008).

Hoplias malabaricus is a Neotropical fish with ecological and commercial importance, and has a wide geographical distribution, occurring in almost all the hydrographic basins of South America. It is considered an opportunistic species since changes in diet due to food supply and competition were detected. Besides, it is considered a keystone species, structuring fish assemblages in isolated lakes (Lowe-McConnell, 1975; Prado et al., 2006; Luz-Agostinho et al., 2006; Novaes and Carvalho, 2011).

H. malabaricus is of commercial importance and is also used by the fishing communities as a food supplement, due to the abundance of this species in freshwater ecosystems of the semi-arid region of Northeastern Brazil. This work aims to describe the feeding habits besides anatomical and histological characterization of the digestive tract of *H. malabaricus*.

Materials and Methods

Study Site and Fish Sampling

Fish sampling was carried out on a monthly basis from July, 2011 to June, 2012, in the Marechal Dutra Reservoir, State of Rio Grande do Norte, located in the Caatinga ecoregion of Northeastern Brazil (6°26'11''S and 36°36'17''W). This reservoir was constructed on River Acauã in the hydrographic basin of River Piranhas-Assu, which is the main hydrographic basin of Rio Grande do Norte, covering an area of 44,000 km² (Chellappa et al., 2009). Fish samples were captured with the help of local fishermen, who used gill nets with different mesh sizes (varying from 6 to 12 cm) with a 12-h fishing effort per month.

Measurements

Fish captured were numbered, measured, weighed and samples of fish were used for morphometric measurements to confirm the taxonomical identification of *H. malabaricus*. Measures of total length (Lt±1cm), total body mass (Wt±1 g) and stomach weight (Wg±0.1mg) were recorded. In this study 133 (54 males and 79 females) individuals of the study species were used, with total body length varying from 12.4 to 40.5 cm, (31.7±9.0) cm for males, and from 12.6 to 41.5 cm (33.1±6.7) for females. The body weight varied from 18.0 to 1028.0 g (467.4±265.2) for males, and from 20.0 to 1000.5 g (484.6±218.8), for females.

Procedures

The digestive tract was removed from the coelomic cavity, using 15 adult samples of *H. malabaricus* for the characterization of the morphology of the digestive tract. The position of the mouth, type of dentition, gills rakers, esophagus, stomach, presence or absence of pyloric caeca and intestine were examined for the morphological description.

For histological descriptions, the digestive tract of 15 adult individuals of *H. malabaricus*, were separated out and fixed in 10% formaldehyde. After 24 hours of fixation, the digestive tract was submitted to microtomy to obtain fragments of the digestive organs: esophagus, stomach (cardiac, fundic, and pyloric regions), pyloric caeca, and the intestine. Tissue samples were fixed in Bouin's fluid or buffered formaldehyde. The fragments were submitted to routine histological techniques (dehydration, diaphanization, and paraffin embedding) and microsectioned at 5 μm. Samples were stained with haematoxylin-eosin (HE) and periodic acid Schiff (PAS) following routine procedures (Smith, 1996).

The esophagus, stomach (cardiac, fundic, and pyloric regions), pyloric caeca, and the intestine were analyzed through transverse and longitudinal sections.

The Intestinal Coefficient

The intestinal coefficient (IC = length of the intestine/total fish length) was calculated (Bertin, 1958). Intestinal lengths (cm) and body lengths (cm) were obtained from 133 specimens of pooled samples of males and females of *H. malabaricus*.

Stomach Contents Analyses and Statistical Analyses

The stomach contents of 133 specimens of *H. malabaricus* were examined under a stereoscopic microscope, and identified to the smallest taxonomic level possible (Holthuis,

1952). The frequencies of occurrence (FO) were calculated (Hynes, 1950; Hyslop, 1980). Index of Alimentary Importance (IAI) was applied on the values of frequency of occurrence to provide the food items that contribute most to the diet of the species (Kawakami and Vazzoler, 1980). Degree of Fullness of Stomachs was estimated (Berg, 1979; Montenegro et al., 2012). The degrees of stomach fullness varied from 0 to 3, where: 0 = empty; 1 = up to 25% full; 2 = between 25% and 75% full; 3 = between 75% and 100% full. The food preference was analyzed using the chi-square (χ^2) test, in order to identify statistically the significant differences between the most consumed item with the others.

RESULTS

Gross Anatomy

H. malabaricus has a large head with a wide dorso-terminal mouth with lips extending to the sides of the head. The teeth are uneven, strong and sharp (Figure 1A and B). Each gill chamber has four gill arches, the first gill arch possesses well spaced seven gill rakers which are short and sharp while the other arches do not exhibit well developed gill rakers (Figure 1C).

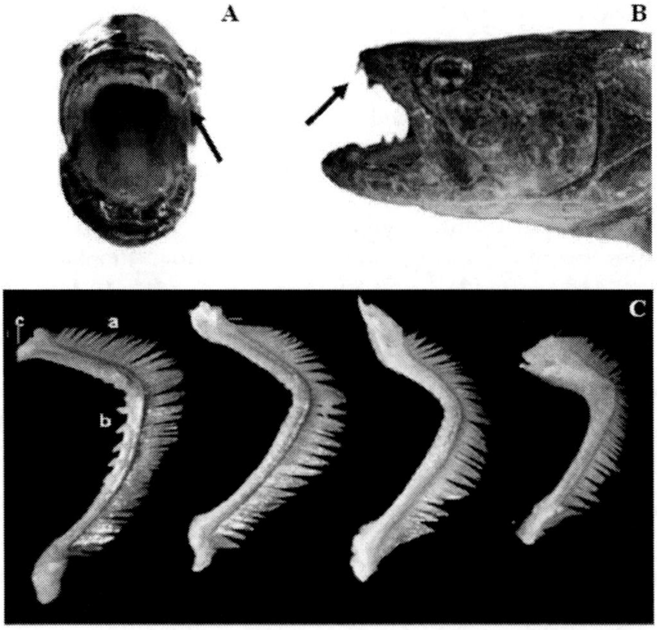

Figure 1. (A) Frontal view and (B) lateral view of the head of *Hoplias malabaricus*, showing the canine teeth (arrows); (C) The branchial arches (a) branchial filaments, (b) gill rakers, (c) bony structure.

The esophagus of *H. malabaricus* is always covered by the beginning of the large intestine. The esophagus is a dorsally flattened short tube, with great elasticity. The stomach is rectilinear in shape with three distinct regions the cardiac, fundic and pyloric. The cardiac region is the longest, located just after the esophagus, with ample light and is tubular in shape.

The fundic region is straight and short with a blind end, which begins the pyloric region which is also short and tubular. The pyloric region extends towards the beginning of the intestinal tube. At the junction of the stomach with the intestine, there are the pyloric caeca, which are short, ramified and are in great numbers. On an average, there are about 113.3±DP 12.7 caeca, distributed in two groups, one on the left side of the small intestine and another on the right side (Figure 2 A and B).

The intestine is short, with an estimated mean intestinal coefficient of 0.72±0.09, and could be subdivided into proximal and distal intestine, in relation to the stomach. The entire region covered by pyloric caeca may be considered as the small intestine, after which comes the distal intestine with the absence of caeca (Figure 2B).

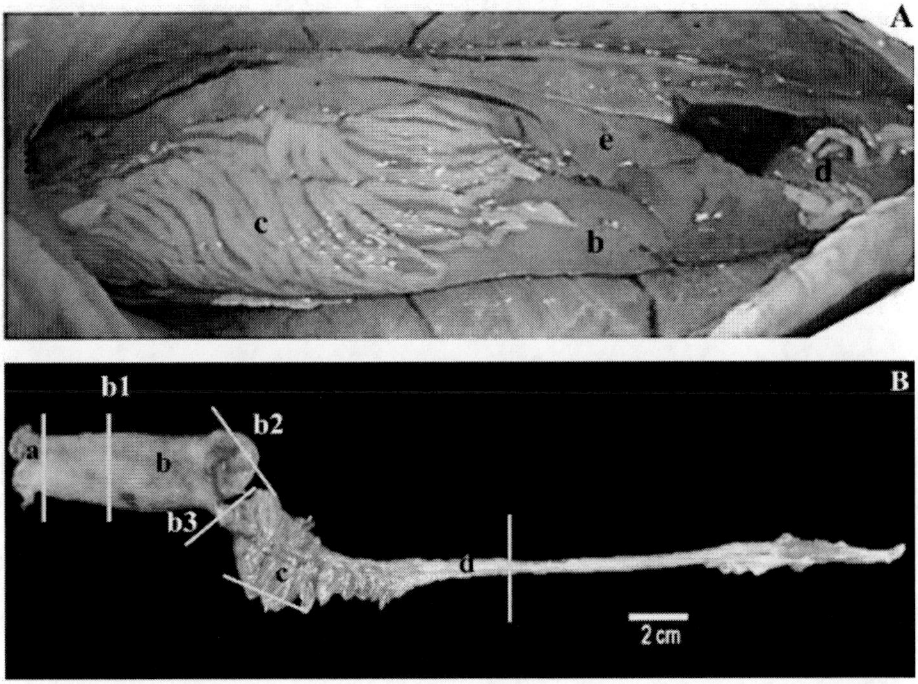

Figure 2. (A) View of the coelomic cavity and (B) structures of the digestive tract of *H. malabaricus*: (a) esophagus, (b) stomach, (b1) cardiac stomach, (b2) fundic stomach, (b3) pyloric stomach, (c) pyloric caeca, (d) intestine and (e) liver.

Histology of the Digestive Tract

The histological analyses evidenced, along the entire digestive tract, the presence of four distinct tunicae composing the wall: mucosa (epithelium, lamina propria, muscularis mucosae), submucosa (submucosal plexus), muscularis (myenteric plexus) and serosa. Besides there is distinct intrinsic innervations represented by the submucosal and myenteric ganglionated plexus, in accordance to the organ considered.

Esophagus and Stomach

The esophagus is covered by a layer adventitia, while the esophageal wall is thick and very distensible. When approaching the stomach, the esophageal diameter becomes reduced, and the insertion of the pneumatic duct in the dorsal region of the left side of the esophagus indicates the beginning of the cardiac region. In the transition region between the two organs, the epithelium becomes cylindrical, decreasing the secretory activity, and the beginning of the gastric glands (Figure 3A and B).

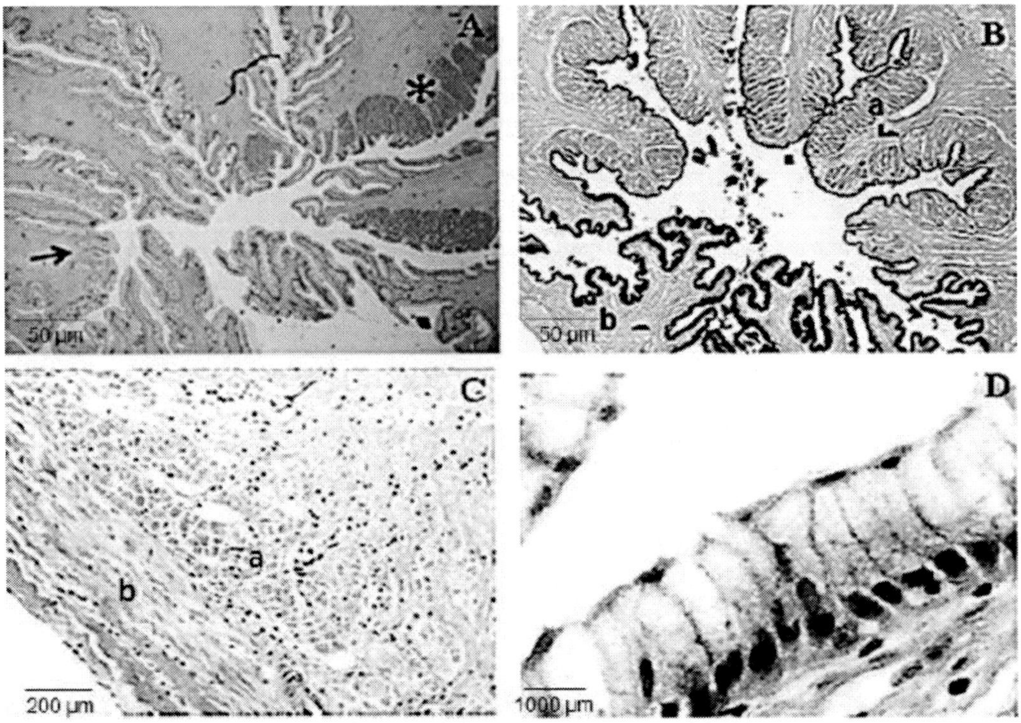

Figure 3. Transition between esophagus and stomach of *H. malabaricus*. (A) esophagus region without gastric glands (arrow) and gastric glands in the cardiac region of the stomach (*); (B) (a) esophageal ramified folds and (b) rounded stomach folds; (C) esophagus: (a) longitudinal muscular layer, (b) circular muscular layer; (D) esophageal pseudostratified mucosecretory epithelium.

The lamina propria, is constituted of continuous loose connective tissue, without glands and merges with the muscle. In this, acidophilic granular cells and some dispersed leukocytes can occur, however, adipose cells were not evident. Muscle fibers, mostly striated, were observed longitudinally arranged in a transversal beam, soon after the lamina propria. The circular muscle fibers are more developed than the longitudinal ones (Figure 3C and D). Next comes the circular muscle fibers, predominantly smooth, and as they approach the cardiac region, gradually assume internal position in the muscular tunica (Figure 4A and B).

The entire stomach is lined by simple columnar epithelium consisting solely of secretary cells which exhibit apical positive PAS (Figure 5C). These cells are uniformly columnar with basophilic nucleus, elliptical, usually located in the middle region of the cell, having homogeneous acidophilic cytoplasm to HE (Figure 4C).

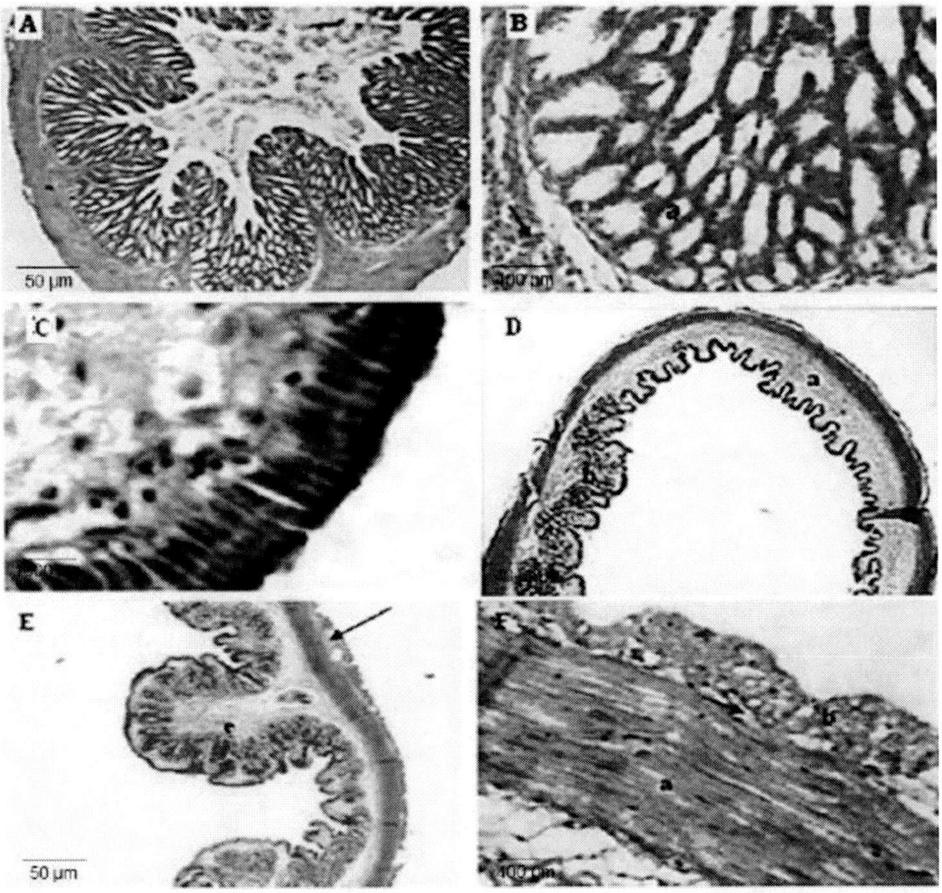

Figure 4. Stomach of *H. malabaricus*: (A) primary cardiac folds; (B) (a) ramified tubular stomach glands, submucosa with acidophylic granular cells (arrow); (C) simple cylindrical epithelium near the gastric crypts; (D) fundic region of the stomach (a) region without folds, (b) region with folds; (E) posterior cardiac tunica (e) esophageal fold of small size, myenteric plexus (arrow); (F) (a) circular muscular tunica, (b) longitudinal muscular tunica, myenteric plexus (arrow).

They can vary in size becoming larger in apices of the mucosal folds, with nuclei displaced to the apical region, and smaller to form the gastric crypts. The lamina propria consists of loose connective tissue which is similar to the walls of the glandular cardiac and fundic regions, and non glandular in the portion before the pyloric region. These exocrine glands are tubular, simple or branched, containing many zymogen granules distributed throughout the cytoplasm.

The muscularis mucosa was not observed, being distinguished from the submucosal lamina propria due to presence of folds of the mucosa and richly vascularized, containing many granular acidophilic cells and leukocytes depending on the stomach region (Figure 4D). The muscular tunica layer, composed of internal circular and external longitudinal layers, presents a well developed myenteric plexus (Figure 4E and F). This plexus exists between the longitudinal and circular layers of muscularis externa in the gastrointestinal tract. Following is the serous tunic that has a sub-serosa containing many blood vessels and nerves.

In the cardiac stomach, the mucosa is thick and forms three to seven primary folds, which are high and rounded along the lamina propria. These folds as they approach the fundic region become smaller or disappear, however, the posterior cardiac region has lower folds (Figure 5A and B). The lamina propria shows tubular glands that open directly into the narrow and shallow gastric pits, those which constitute isolated groups surrounded by inter-glandular septa, derived from the underlying connective tissue that makes the submucosa.

The submucosa is thick and rich in blood vessels, constituted by loose connective tissue with many leukocytes and isolated granular acidophilic cells (Figure 5D). Smooth muscle fibers when present depart from the circular muscle layer. The tunica muscular features thick circular muscle than the longitudinal layer which always remains thin. Among these layers, as well as in the subserosa, occur large nerves and blood vessels of various sizes. The fundic stomach folds present small primary folds, similar to the cardiac region, but no secondary folds occur between them and the gastric pits in this region are shallow (Figure 5).

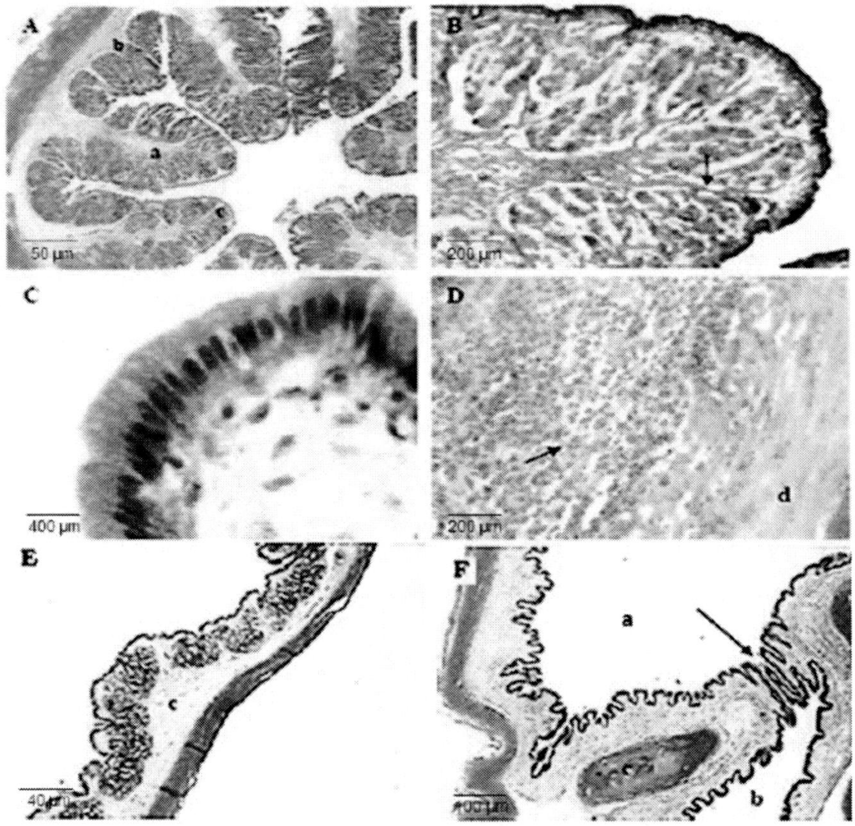

Figure 5. Cardiac region of the stomach of *H. malabaricus* (A) anterior cardiac (a) primary folds, (b) secondary folds, (c) gastric glands; (B) primary stomach fold, interglandular septum (arrow); (C) simple cylindrical epithelium; (D) Submucosa containing many acidophylic granular cells (arrow), (d) circular muscular layer; (E) non-glandular submucosa of the gastric fold; (F) pyloric region of the stomach (a) stomach lumen, (b) intestinal lumen, (c) pyloric sphincter, pyloric valve (arrow).

The lamina propria and submucosa show the same characteristics of the cardiac region, however, glands in fundic region are longer and branched. The tunica muscular layer follows the same constitution as described for cardiac region, however, the myenteric plexus and submucosa have small nerves and small blood vessels.

In the portion near the beginning of the pyloric stomach, the mucosa in some specimens, fails to show the primary folds and the lamina propria becomes without glands, with numerous gastric crypts and submucosa with increased amount of consistent collagen fibers. The pyloric stomach presents numerous parallel and very narrow longitudinal folds, filling the entire lumen of the organ till the pyloric valve, where they become shorter and simplified. At the distal end of the pyloric region appears the pyloric sphincter. In addition to the pyloric sphincter, the pyloric valve is also formed by the folds of the mucosa, which show an elongation towards the intestinal lumen (Figure 5E and F).

The epithelium of the entire stomach, presents little secretory activity, evidenced by PAS and intensity similar to the cardiac and fundic regions. The mucosal lamina propria consists of loose non glandular connective tissue and continuing with the túnica muscular layer. This tunica has inner circular layer four to five times thicker than the longitudinal. The myenteric plexus lacks the big nerves that occurs in the subserosa, but has blood vessels of various sizes.

Intestine

The wall of intestine consists of mucosa, lamina propria, submucosa, muscularis and serosa. Presents numerous mucosal folds in network format, oblique, transverse and longitudinal anastomose. The intestinal epithelium is simple columnar type with goblet cells scattered among the enterocytes (absorptive cells). In goblet cells the whole cytoplasm reacts strongly to PAS, but the enterocytes exhibit no change in their color when subjected to this method.

The muscularis consis of smooth muscles were arranged in a thick circular inner layer and a thin longitudinal outer layer. Between these two muscular layers there were many nerve plexus. The tunica muscular is very thin, and follows the structural pattern found in the stomach, consisting of a of inner circular layer and outer longitudinal layer, both of smooth muscle fibers, and between these layers the myenteric plexus is formed (Figure 6B). The lamina propria is non glandular, and is of loose connective tissue, very irrigated with blood vessels and at some points of the large intestine there are lymphatic nodes (Figure 6C). In the small intestine, the circular layer has approximately twice the thickness of the longitudinal muscle layer. In large intestine the thickness of the two layers is similar.

Pyloric Caeca

These have structure similar to the small intestine, excepting that the light is narrower, thinner tunica muscular layer, and with smaller number of goblet cells in the mucosa (Figure 6D).

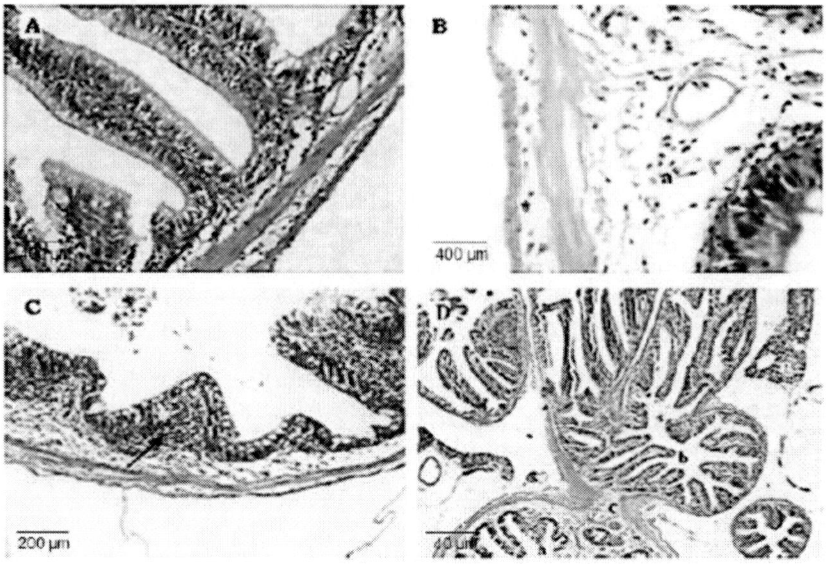

Figure 6. Intestine and pyloric caeca of *H. malabaricus* (A) Intestinal folds; (B) (a) aglandular submucosa, myenteric plexus (*); (C) accumulated lymph in the Intestine (arrow); (D) Insertion of the ramified pyloric caeca, (a) lumen of the intestine, (b) lumen of the caeca, (c) origin of caeca from the intestine.

Feeding Habits

The Index of Alimentary Importance (IAI) indicates that *H. malabaricus* feeds on animal matter, out of which preferentially 67.8% was fish ($\chi^2_{calc.}$ = 3,84; g.l = 1; p<0.05), such as *Oreochromis niloticus* (25.3%), *Plagiosion squamosissimus* (22.1%), *Leporinus piau* (11.9%), *Triportheus angulatus* (8.5%), followed by 21.5% of prawns, *Macrobrachium amazonicum*, and 10.7% of un identified semi-digested food material, characterizing it as a carnivore (Figure 7A).

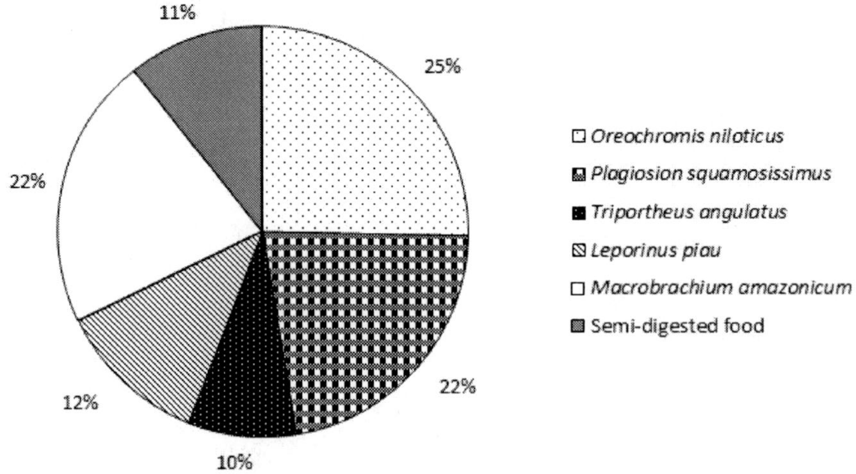

A

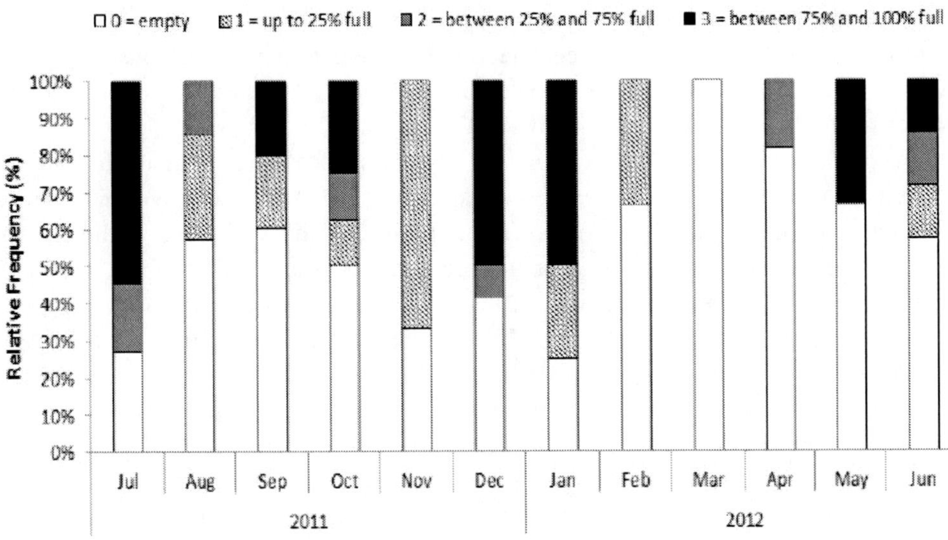

B

Figure 7. (A) Food items encountered in the stomach; (B) Stomach fullness Index (GR) of *H. malabaricus*.

Degree of Fullness of stomachs shows that 25.3% of the individuals had full stomachs, 10.1% partially full, 11.4% partially empty and most of them 53.3% had empty stomachs. The highest percentage of full stomachs was observed during the month of July, while during the month of March a higher percentage of empty stomachs was observed (Figure 7B).

Discussion

The position of the mouth, type of dentition, degree of mouth protractility and the type of the gill rakers are related to the feeding habits of fish (Motta, 1984; Albrecht et al., 2001). *H. malabaricus* in the adult stage is a voracious carnivore, preferentially piscivorous. The type of mouth is characteristic of predatory forms, allowing it to grab and eat a whole large prey. In some carnivorous fish, the gill rakers are modified to grab and prevent the escape of the prey, or to scrape and grind the food. The strong teeth, short and sharp gill rakers are anatomical adaptations shared between several species of psivores of Characiformes (Mittelbach, 2002; Peretti and Adrian, 2008).

The organization of esophageal longitudinal folds is probably related to the distensibility of the walls of this organ, adapting it for the conduction of food to the stomach (Montenegro et al., 2012). *H. malabaricus* has many folds in the constitution of the esophagus to facilitate the ingestion of prey whole.

Ingestion of large prey by *H. malabaricus* is also made possible due to the cardiac region of the stomach having long and distensible walls, whose longitudinal folds, in large numbers, extend the gastric lumen. The fundic region being in the form of a sac with small folds, functions as a store house of improperly digested food particles, which are prevented from reaching the intestines due to the small dimension of the pyloric region (Peretti and Adrian,

2008). Cardiac and fundic regions have thick folds which help in storing large volume of digested food. The pyloric region directs the flow of food into the intestine. The pyloric region does not have glands, which was also demonstrated for other carnivorous fish (Petrinec et al., 2005; Carrasson et al., 2006; Montenegro et al., 2012).

The intestine of *H. malabaricus* is short, an anatomical feature similar to that observed for the carnivorous species *Culter alburnus* (Cao et al., 2011). The carnivorous fish have short intestine, with an intestinal coefficient from 0.5 to 2.4 and variations in the length of intestine are offset by changes in the intestinal mucosa (Bertin, 1958; Jobling, 1995). The intestinal coefficient has been investigated in many fish, and has been used as an important criterion for classification of their feeding habits (Albrecht et al., 2001; Moraes et al., 2004; Cao et al., 2011).

The pyloric caeca, when present, may vary in number, shape and structure between species. There are more pyloric caeca in fish with short intestine. In *H. malabaricus*, they are numerous with the presence of absorptive cells and goblet cells, which suggests the involvement of the cecca with the absorption of nutrients, and lubrication of the intestinal epithelium. The presence of pyloric caeca is considered a strategy evolved to increase the intestinal surface of absorption without increasing the length of the intestine (Cinar and Senol, 2006).

The number of empty stomachs observed during the study period, could be explained by the carnivorous diet of this fiah. A diet based on fish has high energetic value, thereby reducing the need for continuous intake of food. This also indicates resistance to long periods of fasting, which was also found for other species of carnivorous fish (Nikolsky, 1963; Alvim and Peret, 2004; Lowrya et al., 2005; Montenegro et al., 2012). Histology of the digestive tract of *H. malabaricus* confirms its carnivorous habits of feeding refelected by the stomach content analyses.

ACKNOWLEDGMENTS

Thanks are due to the National Council for Scientific and Technological Development of the Ministry of Science and Technology of Brazil (CNPq/ MCT) for the Research grants offered to the authors.

REFERENCES

Abdulhadi, H. A. A. 2005. Some comparative histological studies on alimentary tract of tilapia fish (*Tilapia spilurus*) and sea bram (*Mylio cuvieri*). *Egyptian Journal of Aquatic Research*, 31: 387-397.

Albrecht, M. P., Ferreira, M. F. N. and Caramaschi, E. P. 2001. Anatomical features and histology of the digestive tract of two related neotropical omnivorous fishes (Characiformes; Anostomidae). *Journal of Fish Biology*, 58: 419–430.

Alvim, M. C. C. and Peret, A. C. 2004. Food resources sustaining the fish fauna in a section of the upper Sao Francisco River in Três Marias, MG, Brazil. *Brazilian Journal of Biology*, 64: 195-202.

Berg, J. 1979. Discussion of methods of investigating the food of fishes, with reference to a preliminary study of the prey of *Gobiusculus flavescens* (Gobiidae). *Marine Biology*, 50: 263-273.

Bértin, L. 1958. Appareil digestif. In: GRASSÉ, P.P. (Ed.). *Traité de zoologie*. Paris: Masson, 13: 1249-1301.

Canan, B., Pessoa, E. K. R., Volpato, G. L., Araujo, A. and Chellappa, S. 2011. Feeding and reproductive dynamics of the damselfish, *Stegastes fuscus* in the coastal reefs of northeastern Brazil. *Animal Biology Journal*, 2: 113-126.

Canan, B.; Nascimento, W. S.; Silva, N. B. and Chellappa, S. 2012. Morphohistology of the digestive tract of the damsel fish, *Stegastes fuscus* (Osteichthyes: Pomacentridae). *The Scientific World Journal*, 2012: 1-9.

Cao, X. J., Wang, W. M. and Song F. 2011. Anatomical and Histological Characteristics of the Intestine of the Topmouth Culter (*Culter alburnus*). *Anatomia, Histologia Embryologia*, 40, 292–298.

Carrasson, M., Grau, A., Dopazo, L. R. and Crespo, S. 2006. A histological histochemical and ultrastructural study of the digestive tract of *Dentex dentex* (Pisces, Sparidae). *Histology and Histopathology*, 21: 579-593.

Cinar, K. and Senol, N. 2006. Histological and histochemical characterization of the mucosa of the digestive tract in tlower fish (*Pselluophoxinlls antalyae*). *Anatomia, Histologia Embryologia*, 35: 147-151.

Holthuis, L. B. 1952. A general revision of the Palaemonidae (Crustacea: Decapoda Natantia) of the Américas. II – The subfamily Palaemoninae. 23p.

Jobling, M. 1995. Environmental Biology of Fishes. London: Chapman & Hall.

Kawakami, E. and Vazzoler, G. 1980. Método gráfico e estimativa de índice alimentar aplicado ao estudo de alimentação de peixes. *Boletim do Instituto Oceanográfico*, 29: 205-207.

Kapoor, B. G. 1975. Anatomy and histology of the alimentary canal in relation to its feeding habits of a siluroid fish, *Wallago attu*. *Journal of the Zoological Society*, 5: 191-210.

Lowe-Mcconnell, R. H. 1975. *Fish Communities in Tropical Freshwaters: Their Distribution, Ecology and Evolution*. London and New York: Longman. 283p.

Lowrya, D., Wintzera, A. P., Matotta, M. P., Whitenacka, L. B, Huber, D. R., Dean, M. and Motta, P. J. 2005. Aerial and aquatic feeding in the silver arawana, *Osteoglossum bicirrhosum*. *Environmental Biology of Fishes*, 73: 453-462.

Luz-Agostinho, K. D. G., Bini, L. M., Fugi, R., Agostinho, A. A. and Júlio JR., H. F. 2006. Food spectrum and trophic structure of the ichthyofauna of Corumbá reservoir, Paraná river basin, Brazil. *Neotropical Ichthyology*, 4: 61-68.

Mittelbach, G. G. 2002. Fish foraging and habitat choice: a theoretical perspective. In Hart, P. J. B. and Reynolds, J. D. (Ed.). *Fish Biology and Fisheries.*

Montenegro, L. A., Silva, N. B., Nascimento, W. S. and Chellappa, S. 2012. Anatomy and Histology of the digestive tract and feeding habits of the marbled swamp eel *Synbranchus marmoratus*. *Animal Biology Journal*, 3 (3): 127-143.

Moraes, M. F.; Freitas-Barbola, I. and Duboc, L. F. 2004. Feeding habits and morphometry of digestive tracts of *Geophagus brasiliensis* (Osteichthyes, Cichlidae), in a lagoon of high tibagi river, Parana state, Brazil. *Biological and Health Sciences*, 10: 37-45.

Motta, P. J. 1984. Mechanics and functions of jaw protrusion in Teleost fishes: a review. *Copeia*, 1: 1-18.

Nikolsky, G. V. 1963. *The ecology of fishes*. Academic Press. London. 352p.

Novaes, J. L. C. and Carvalho, E. D. 2011. Population structure and stock assessment of *Hoplias malabaricus* (Characiformes: Erythrinidae) caught by artisanal fishermen in river-reservoir transition area in Brazil. *Revista de Biología Tropical*, 59: 71-83.

Novakowski, G. C.; Fugi, R. and Hahn, N. S. 2004. Diet and dental development of three species of *Roeboides* (Characiformes: Characidae). *Neotropical Ichthyology*, 2: 157-162.

Peretti, D. and Andrian, I. F. 2008. Feeding and morphological analysis of the digestive tract of four species of fish (*Astyanax altiparanae*, *Parauchenipterus galeatus*, *Serrasalmus marginatus* and *Hoplias malabaricus*) from the upper Paraná River floodplain, Brazil. *Brazilian Journal of Biology*, São Carlos, SP, 68: 671-679.

Petrinec, Z., Nejedli, S. Kuir, S. and Opaëak, A. 2005. Mucosubstances of the digestive tract mucosa in northern pike (*Esox lucius* L.) and European catfish (*Silurus glanis* L.). *Veterinarski Arhiv*, 75: 317-327.

Prado, C. P. A., Gomeiro, L. M. and Froehlich, O. 2006. Spawning and parental care in *Hoplias malabaricus* (Teleostei, Characiformes, Erytrinidae) in the Southern pantanal, Brazil. *Brazilian Journal of Biology*, 66: 697-702.

Reis, R. E., Kullander, S. O. and Ferraris-JR., C.J. 2003. *Check list of the freshwater fishes of South and Central America.* EDIPUCRS, Porto Alegre, 729p.

Smith, M. P., Sansom, I. J. and Repetski, J. E. 1996. Histology of the first fish. *Nature* 380: 702-704.

Wootton, R.J. 1999. *Ecology of teleost fish.* The Netherlands: Kluwer Academic Publishers, 386p.

In: Biology of Semiarid Tropical Fish
Editors: J. R. Gomes and S. Chellappa
ISBN: 978-1-63117-067-6

Chapter 8

FEEDING AND REPRODUCTIVE DYNAMICS OF THE DAMSELFISH, *STEGASTES FUSCUS* IN THE COASTAL REEFS OF NORTHEASTERN BRAZIL

Bhaskara Canan[1], Emilly Kataline Rodrigues Pessoa[2], Gilson Luiz Volpato[3], Arrilton Araújo[1] and Sathyabama Chellappa[1*]

[1]Post-Graduate Programme in Psychobiology, Centro de Biociências, UFRN, Campus Universitário, Natal, RN, CEP: Brazil
[2]Department of Oceanography and Limnology, Universidade Federal do Rio Grande do Norte, Praia de Mãe Luiza, s/n, Natal, RN, CEP: Brazil
[3]Laboratório de Fisiologia e Comportamento Animal, Departamento de Fisiologia, IB, CAUNESP, UNESP, Botucatu, SP, CEP: Brazil

ABSTRACT

This study reports on the temporal dynamics of feeding and reproduction of the damselfish, *Stegastes fuscus* in the coastal rocky reefs of northeastern Brazil. Water temperatures and rainfall data were registered and fish were captured on a monthly basis during one year. Fish body weights and lengths were measured and the stomachs were removed and classified according to the degree of fullness. The gonads were weighed, examined for sex determination and maturation was determined based on macroscopic inspections. A higher frequency of females (78%) was registered in relation to males (22%). The lowest mean degree of stomach fullness was observed in August and the highest in January. Only 2.49% of the fish had empty stomachs and 97.51% had mainly macroalgae in their stomachs. The period of February to August was associated to a long phase of gonadal resting in males and females. Two peaks of spawning were registered during January and September - October.

[*] E-mail of the corresponding author: chellappa.sathyabama63@gmail.com.

Keywords: Pomacentridae, damselfish *Stegastes fuscus*, feeding, reproduction, environmental parameters

INTRODUCTION

The species that inhabit coral reefs are strongly influenced at low tides by external factors such as temperature, luminosity, and rainfall, among others (Klumpp *et al.* 1987; Cleveland and Motgomery 2003). These effects are due to the low volume of water in these areas, which are usually trapped in tide pools. Thus, it is expected that tropical fish communities living in tide pools are influenced by external factors which vary throughout the year. Marine fish populations show a wide adaptive radiation in foraging habits and exhibit annual changes in their feeding pattern (Lison de Loma *et al.*, 2000) and reproductive dynamics (Richardson *et al.*, 1997; Araújo and Chellappa, 2002; Asoh 2003; Picciulin *et al.*, 2004). Effects of temperature, luminosity and rainfall on annual distribution and reproduction have been detected in marine fish (Gibson *et al.*, 1993; Chellappa et al., 2010). However, few studies have dealt with tropical reef fish (Schwamborn and Ferreira, 2002; Osório *et al.*, 2006).

Fish in the equatorial regions are more influenced by rainfall and are less subjected to variations in temperature and photoperiod, since they are practically constant all around the year (Volpato and Trajano 2006; Chellappa et al., 2010). Thus, fish in the equatorial regions are useful for investigating the influence of the dry and rainy periods on their biology, since other intervening factors are rather reduced.

The damselfish, *Stegastes fuscus* (Cuvier 1830), (Osteichthyes: Perciformes: Pomacentridae), occurs in tide pools located in the coastal rocky reefs in northeastern Brazil was investigated. The objective of this study was to verify the sex ratio, the body length and body mass structure of males and females, the degree of stomach fullness and gonadal development of *S. fuscus* in order to understand the temporal dynamics of feeding and reproduction of this species. An attempt was made to correlate the spawning season of *S. fuscus* with rainfall, one of the environmental factors known to modulate the duration and timing of the spawning period of tropical fish.

MATERIALS AND METHODS

Study Area and Sample Collection

Samples of damselfish *S. fuscus* were captured from the tide pools located in reef formations (approximately 3 km in extension), in Búzios Beach, Nizia Floresta, Brazil (06º00'40''S; 35º 06'38''W), located in the South Atlantic Equatorial Zone (Figure 1). In the collection site there were other co-occurring fish species pertaining to Pomacentridae, Serranidae, Lutjanidae, Haemulidae and Scianidae. The benthonic macroalgal flora of this area include Rhodophyta, Chlorophyta and Ochrophyta with dominance of *Hypnea musciformis* (Rhodophyta), *Gracilaria* spp (Rhodophyta), *Solieria filiformis* (Rhodophyta), *Gelidium coarctatum* (Rhodophyta) and *Padina gymnospora* (Ochrophyta) (Marinho-Soriano et al., 2009).

Fish were captured on a monthly basis during the lowest daytime tides, from September 2005 to August 2006. The observations on fish occurrence were made at low tide periods, when they were influenced by variations during the course of the year. Water temperature (°C) was recorded monthly in the field when the fish were captured. Rainfall data (mm) was obtained from the Meteorological Station of the Universidade Federal do Rio Grande do Norte (UFRN), Natal, Brazil.

Fish were captured with a 0.5 cm mesh hand net, 90 min before and 90 min after the peak low tide. The captured specimens were placed in plastic bags and packed on ice in a thermal container and were then transported to the Laboratory for immediate inspection. The voucher specimens were deposited in the fish collection of the Ichthyology Laboratory of the Departamento de Oceanografia e Limnologia, Universidade Federal do Rio Grande do Norte, Brazil.

Measurements

The total body mass (TW) of each fish was registered to the nearest 0.001 g. The fish were measured to obtain standard (SL) and total (TL) body lengths to the nearest 0.1 cm. The gonads were examined to identify the sex of each individual.

Macroscopic Examination of Gonads and Estimation of GSI

The location and general aspects of the gonads were observed and the stages of reproductive maturity were determined macroscopically according to classification of Asoh (2003). Gonadosomatic index (GSI) was calculated as the ratio between gonad mass and total body mass for females (Wootton et al. 1978).

Estimation of the Degree of Fullness of Stomachs

The stomachs were removed and their degree of fullness were classified according to Berg (1979): 0 (empty), 1 (up to 25% full), 2 (between 25% and 75% full) and 3 (between 75% and 100% full). To estimate these percentages, the food contents were accumulated at the bottom of the stomach and then this filled region was estimated in relation to total stomach volume. Stomach contents were identified using identification manuals for macroalgae and Diatomaceae (Werner, 1977; Reviers, 2006; Marinho-Soriano et al., 2009).

Statistical Analyses

Monthly analysis of frequency of occurrence of males and females was performed to obtain the sex ratio. Mean values were compared by ANOVA in order to assess the variations considering month, sex and seasonality (dry and rainy). Tuckey test was used for testing differences between particular pairs of means. The repletion index was square root transformed. The Goodman's test was used to verify the proportions between the numbers of fish (Goodman 1965), while χ^2 was used to test for differences in the frequency of males and females along the year and during the dry and rainy seasons. Significance was attributed at 5% level.

RESULTS

Rainfall and Water Temperature

During the study period, northeastern Brazil experienced six months of dry season from September, 2005 to February, 2006 and during this season the rainfall varied from a maximum of 44.0 mm to a minimum of 2.0 mm, with a mean rainfall of 18.33 mm (±sd 17.39 mm).

The rainy season extended from March to August, 2006, during which period the rainfall varied from a maximum of 752.0 mm to a minimum of 90.0 mm, with a mean of 307.83 mm (± 274.52 mm). There was a significant difference between the rainfall levels during the dry and rainy seasons ($t = 2.578$; $p=0.0275$).

The water temperature varied from a maximum of 36°C in March, 2006 to a minimum of 25°C in July, 2006. During the dry season the mean water temperature was 32.67°C (± 1.21°C), with a maximum of 34.5°C to a minimum of 31°C. During the rainy season the mean water temperature was 30.58°C (± 4.65°C) with a maximum of 37°C to a minimum of 26°C. There was no significant difference between the water temperature during the dry and rainy seasons ($t = 1.259$; $p=0.2366$).

Total Body Length, Body Mass and Sex Proportion

The frequency of individuals by standard length classes showed a normal distribution. Males and females ranged from 3.0-4.5 to 10.5-12.0 cm in length classes (Figure 2a). There were significant differences in the number of males and females throughout the study period, except for the months of July and August (Figure 2b). There was a higher proportion of females (78%) than males (22%) in the total sample (n= 562) ($\chi^2 = 35.75$).

The length-mass relation was not different for males and females indicating that weight increment is positively allometric in relation to total length. The length and mass was correlated by an exponential curve ($r = 0.9876$ for females and $r = 0.9911$ for males). (Figure 3)

The total length of fish was more distinct between the sexes at the beginning of the dry season (Figure 4a; $F = 0.43$; $p=0.83$). The fish were significantly larger in March and April than during the other months ($F = 2.38$; $p=0.04$) (Figure 4a). However, standard length of fish varied between the sexes on a temporal scale.

In the months of July-August and November- December, the females were significantly larger than the males (Figure 4b; $F = 3.24$; $p=0.007$). The females captured in March-April and July-August had higher mean standard length compared to the other months.

The males captured in March-April were the largest and those captured in November-December were the smallest. Body mass was equally affected in relation to sex and month

(Figure 4c; F = 2.75; p=0.02). The females had greater body mass than the males in January-February, whereas the male body mass was the highest in March-April (Figure 4c).

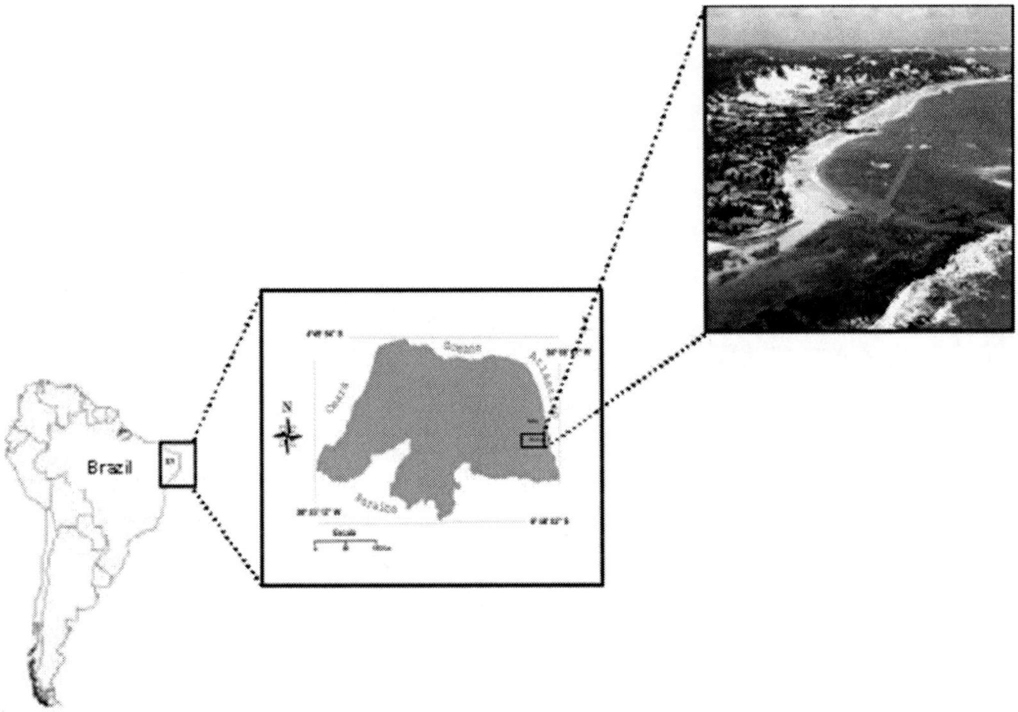

Figure 1. Study area: Búzios Beach, Nísia Floresta, RN, Brazil (The arrow indicates the collection site).

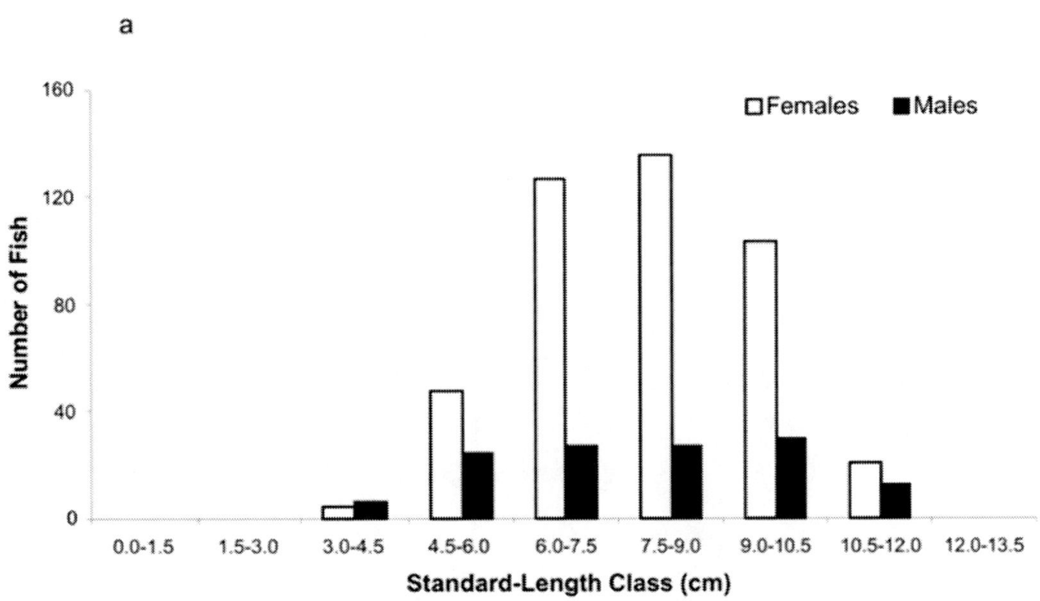

Figure 2. (Continued)

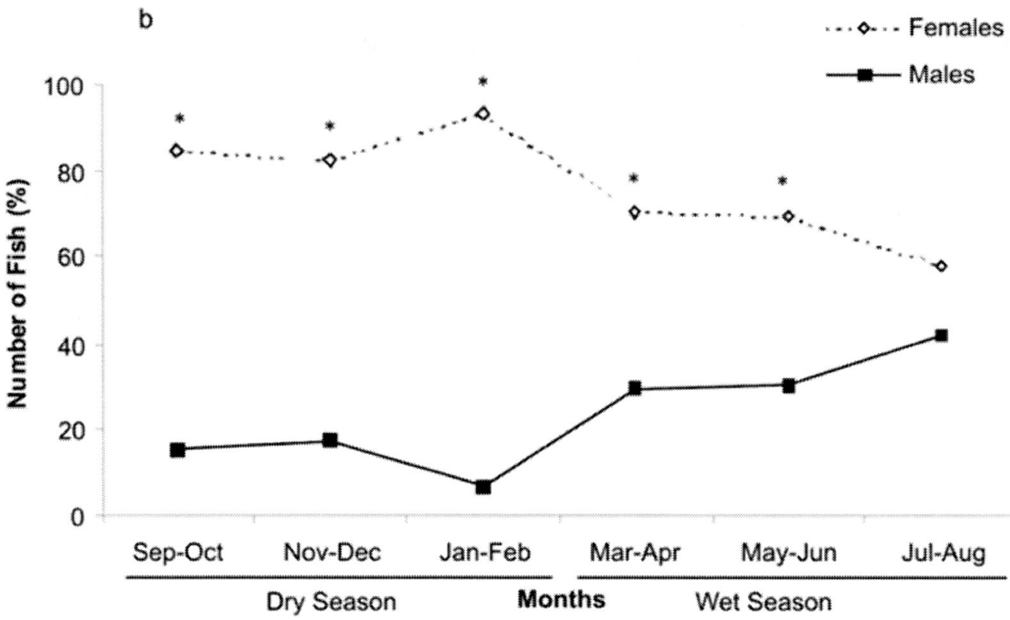

Figure 2. The frequency of occurrence of individuals by standard length classes (a) and monthly variation in percentage of occurrence (b) of males and females of *Stegastes fuscus* in the coastal rocky reefs (The * indicate significant differences).

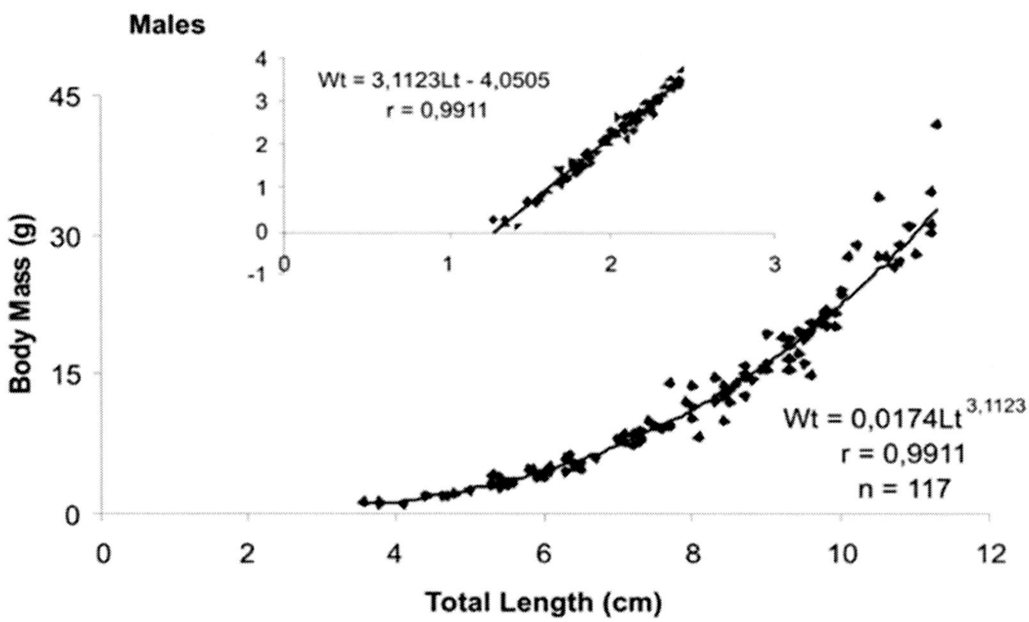

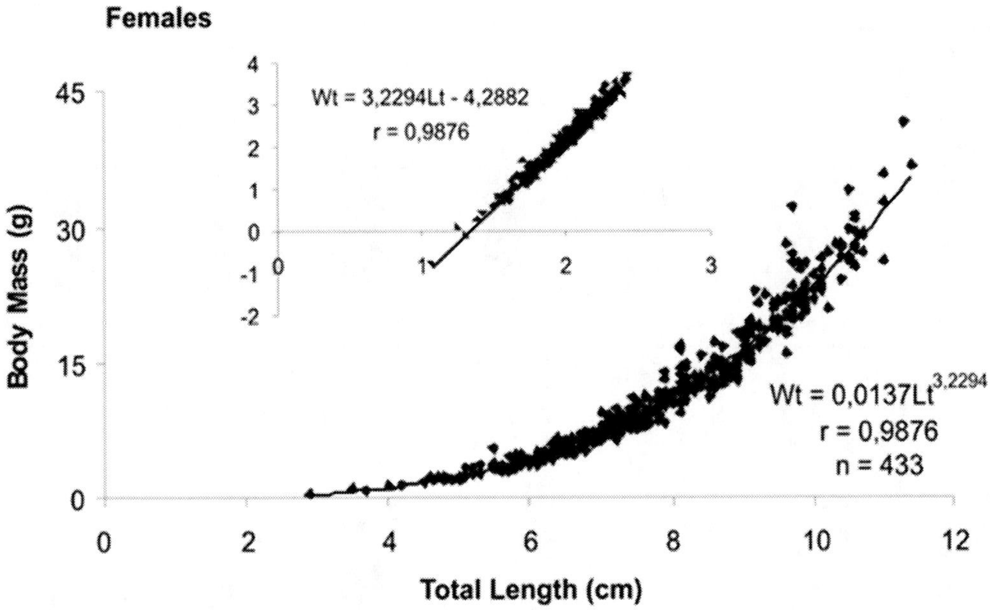

Figure 3. Relation between body length and body mass in males and females of *Stegastes fuscus*.

The Degree of Fullness of Stomachs

Of the 562 specimens examined, a low proportion of 14 had empty stomachs (2.49%), whereas 548 specimens (97.51%) had stomach contents consisting mainly of macroalgae (*Gelidium* sp., *Grasilaria* sp. and *Padina* sp.), benthonic diamatoceae and organic debris. The highest frequencies of empty stomachs occurred from August to December, whereas the highest frequencies of stomachs with food occurred from January to July. The stomach fullness did not vary throughout the year. The lowest mean repletion level (1.59) was in August, and the highest (2.29) occurred in January (Figure 5).

Repletion index of fish varied over the months (Figure 6). No correlation was observed between fish size (standard length, total length and body mass) and repletion index, regardless of sex and season ($r = 0.45$, $p>0.05$, $n = 6$). Repletion index did not differ between males and females ($F = 0.38$; $p=0.54$), despite the variations observed over the months ($F = 2.22$; $p = 0.05$) (Figure 6).

Macroscopic Aspects of Gonads and GSI

Five stages of gonadal maturation (immature, developing, mature, spent and resting) were identified. The gonadosomatic index (GSI) of females was high in September, 2005 and in January, 2006. The period between February and August, 2006 was associated to a gonadal inactivity period. Females with developing and resting gonads predominated in the rainy season (Figure 7a and 8). The highest mean GSI values coincided with the lowest temperatures (Figure 7b).

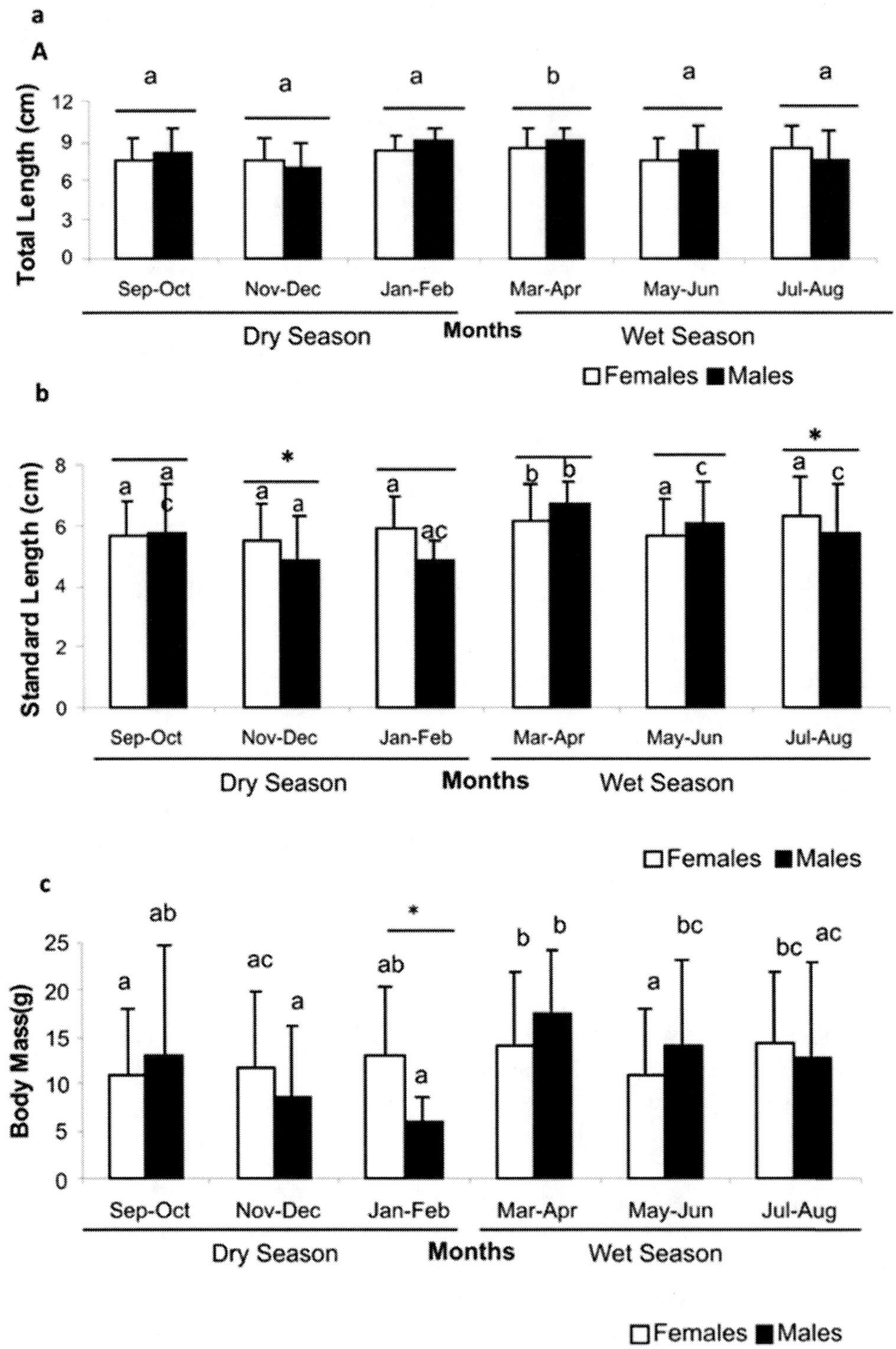

Figure 4. Total length, standard length and body mass of *Stegastes fuscus* in relation to sex and seasonal variation. (The * indicate significant differences between the sex when compared during the same period).

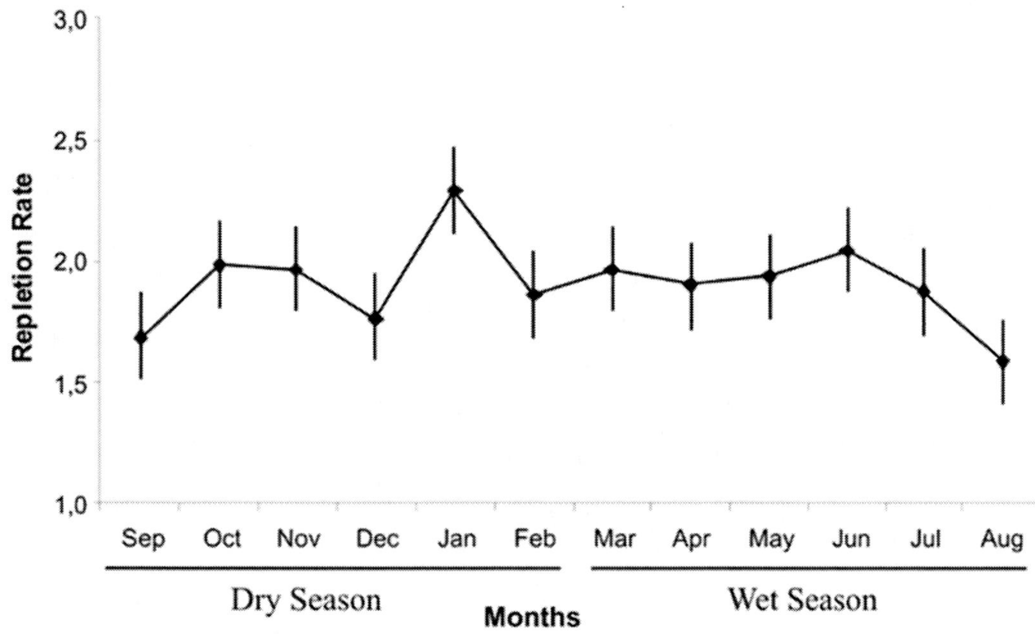

Figure 5. Mean monthly values for degree of stomach fullness for pooled sex in *Stegastes fuscus*.

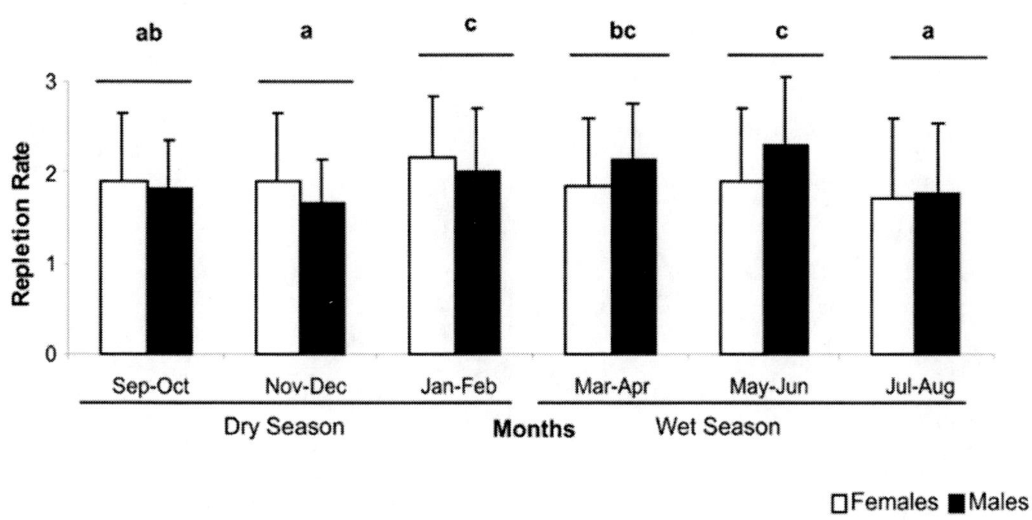

Figure 6. Stomach repletion rate in *Stegastes fuscus* according to sex, season and across months. (The groupings of horizontal lines indicate similar indices and simple letters above the lines indicate significant differences).

Though the development of gonads did not vary much throughout the year, the gonadal development profiles of males and females were different during both dry and rainy seasons (Figure 8). The females were predominately with developing gonads, whereas the males, for the most part, were juvenile and with immature gonads.

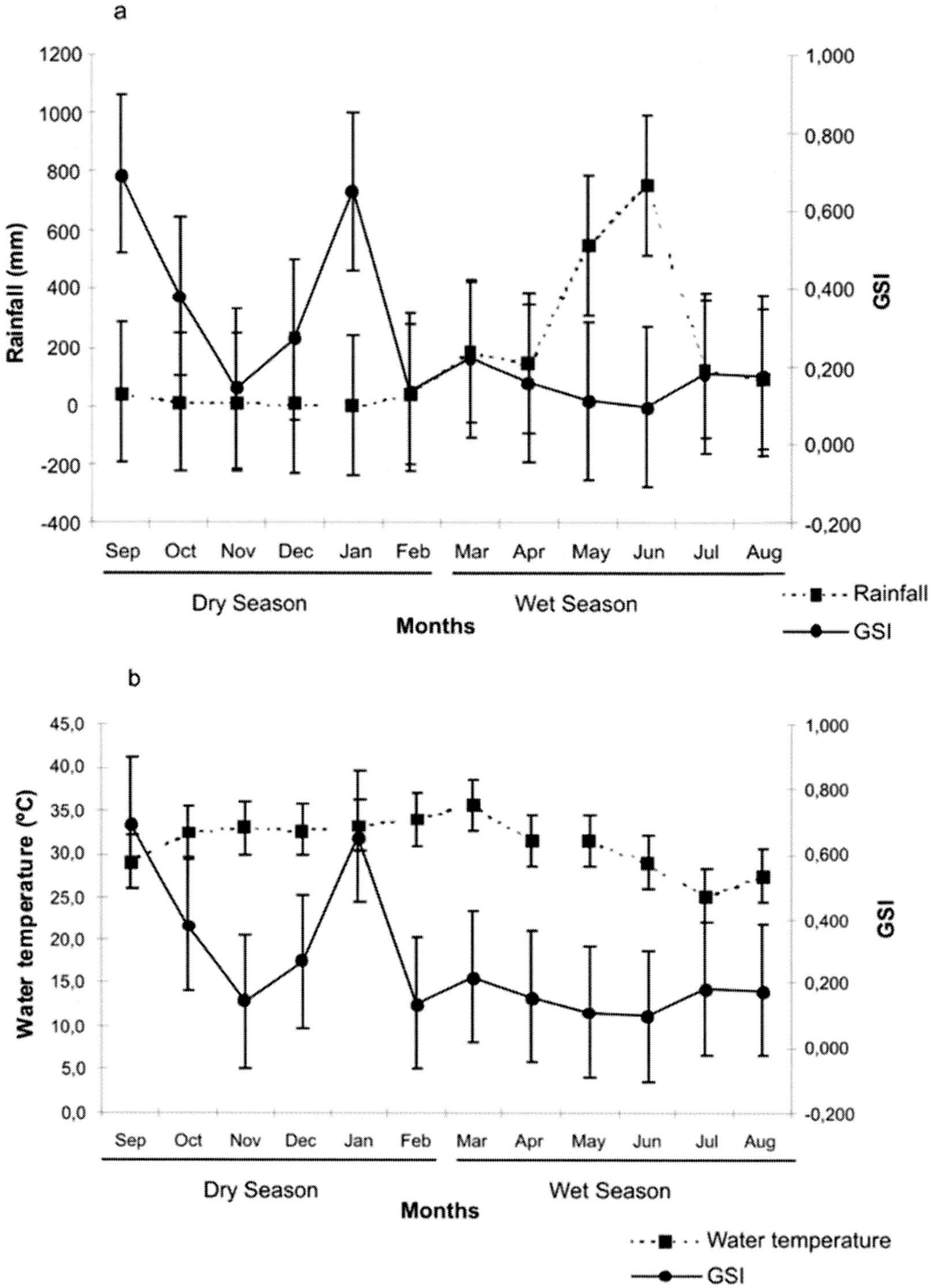

Figure 7. Monthly mean values of (a) rainfall and GSI (b) water temperatures and GSI in females of *Stegastes fuscus*.

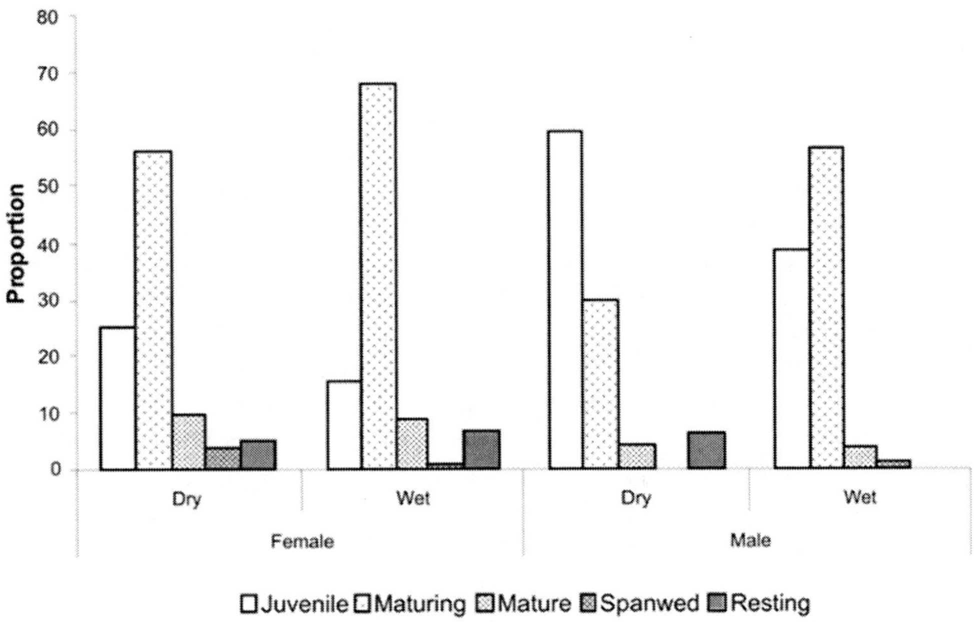

Figure 8. Fish proportion in relation to sex, gonadal status, dry and rainy season in *Stegastes fuscus*.

DISCUSSION

Both males and females of this species showed positive allometric growth. Though the classical sex ratio is a balanced one, usually being 50% of males and 50% of females (Nikolsky 1969), the proportion between males and females was different from 1:1 in the present study, since females were predominant during most of the months. This is possibly due to the behavior of these individuals since territoriality is common among pomacentrids during the reproductive phase. Males of *S. fuscus* guard the eggs in territories before and after they hatch, thus keeping off the predators and unwanted female intruders. Male care also includes ventilation of eggs, cleaning of territory and discarding dead eggs, and these activities occupy about 25% of their time (Petersen 1995). This behaviour, possibly results in potential coasts for the territorial males, which include lack of food during parental care (Tabata 1995; Blazquez *et al.* 1998; Gaillard *et al.* 2004), predation risks from bigger fish and mortality difference between the sexes (Hatcher 1981; Carpenter 1986; Klumpp and Polunin 1989).

The difference in spatial distribution between S *fuscus* males and females, as well as for other species, has been reported (Munro 1976). In this study, S *fuscus* males were distributed predominately along the outer edges of the reefs, which are deeper, while the females were found in shallower areas.

Populational stratification could be another factor responsible for the unbalanced sex ratio of this species. The predominance of males of *S. diencaeus* in Barbados appears to be related to specific exigencies of males and females, resulting from environmental resources, like appropriate sites for breeding territories, proximity to cleaning stations and search for cleaning agents, depending on the parasitic loads (Cheney and Côté 2003). Thus, a possible populational stratification appears to be a more adequate explanation to justify the difference observed in sex ratio and such intra-specific variation is a common characteristic among reef fishes (Nemtzov 1997; Sikkel *et al.* 2005).

Seasonal variation in relation to the body size, food intake and gonadal development of *S. fuscus* was observed, with variations associated to rainfall. The mass-length equation indicates how much one variable changes in association with the variation of the other. In this context, the relation between body mass and length of *S. fuscus* showed slightly higher values compared to the value reported by Bohnsack and Harper (1989). This difference may be explained by the fact that smaller fish (from 3 to 8 cm in furcal length) were used, whereas in the present study the size range of fish was larger (2.9 to 11.4 cm).

Differences in the mass-length relation between sexes have been reported for several fish species (Le Cren 1951), and are a consequence of the higher energy expenditure of females during gonadal maturation (Wootten 1995). Even though the fish in the present study were predominately in the immature and developing gonadal phases, the expected marked effect of sex on the weight-length relation did not occur. One explanation for this finding is that, although female energy expenditure in producing gonads is greater, *S. fuscus* males spend energy defending their territory and protecting the territory (Itzkowits *et al.* 2000; Sikkel *et al.* 2005). Coasts of territorial defense by males may be equivalent to expenditure of energy by female on gonadal development, which could explain the similarity in the mass-length curves of males and females.

This study confirms the herbivorous nature of *S. fuscus* in accordance with Ferreira *et al.* (1998) and Osório *et al.* (2006). There was no difference between the sexes in stomach repletion rates, though there was a higher rate between January and June. The increased ingestion may be related to the reproductive phase. This association may lead to accumulation of sufficient energy to complete the gonadal development. On the other hand, the higher repletion level may result from greater food availability. That is, reproduction may be associated with the period of greater food supply, which would guarantee not only adult development, but also offspring survival. This profile coincides partially with the higher *S. fuscus* length patterns found in March-April and July-August. There was a predominance of larger specimens in the rainy season, which seems to be a response more associated to food availability than to reproduction. The food supply is greater in the rainy season and therefore, in the dry period, the larger specimens must search for other feeding sites, which would explain the size variation of the fish collected in this study.

Considering the gonadal development phase of *S. fuscus,* the females were in a more advanced stage of gonadal development than the males. The action of external influences has been considered as important spawning synchronizers, for example, in *Chromis dispilus* (Tzioumis and Kingsford 1995). The different gonadal development phases between *S. fuscus* males and females suggest that there must be some acceleration mechanism for males or a delay mechanism for females that synchronizes the spawning period in this species. However, rainfall had no effect on overall gonadal development. Since temperature and photoperiod duration are relatively constant throughout the year in the study area, other environmental stimuli might have caused an influence. In tropical regions the dynamics of fish reproduction is associated mainly to rainfall, which is expected for fish populations in an equatorial region (Volpato and Trajano 2006; Chellappa et al., 2010). However, this association was not clear in this study. The spawning of *S. fuscus* occurred mainly in the dry season (September and February), whereas the initial gonadal development stages predominated in the rainy season.

The spawning season observed in this study is slightly different to those reported for pomacentrids from other geographical regions. Sale (1977) reported a 5-9 month spawning season duration for pomacentrids, with multiple, weekly or monthly frequencies, promoting a wide dispersion. In the northeast Caribbean, *Chronis cyanea* spawn in April, *Microspothodon chrysurus* in March, *S. fuscus* in January, June and September, and *S. leucostictus* in September (Erdman 1976). In the Red Sea, the numerous pomacentrid species, with their varied spawning behavior, reproduce from March-April to September (Fischelson et al. 1974).

ACKNOWLEDGMENTS

This study was supported by the National Council for Scientific and Technological Development of Brazil (CNPq) in the form of Research grants (S. Chellappa, A. Araújo and G. L. Volpato) and by the Foundation CAPES/MEC (scholarship awarded to E. K. R. Pessoa).

REFERENCES

Araújo, A.S. and Chellappa, S. (2002). Estratégia reprodutiva do peixe voador, *Hirundichthyes affinis* Gunther (Osteichthyes: Exocoetidae). *Revista Brasileira de Zoologia. 19* (3), 691-703.

Asoh, K. (2003). Gonadal development and infrequent sex change in a population of the humbug damselfish, *Dascyllys aruanus,* in continuous coral-cover habitat. *Marine Biology*, *142* (1), 1207-1218.

Berg, J. (1979). Discussion of methods of investigating the food of fishes, with reference to a preliminary study of the prey of *Gobiusculus flavescens* (Gobiidae). *Marine Biology*, *50* (1), 263-273.

Blazquez, M.S., Zanuy, M.C. and Piferrer, F. (1998). Structural and functional effects of early exposure to estradiol-17 beta and 17 alpha- estradiol on the gonads of gonochoristc teleost *Dicentrarchus labrax. Fish Physiology and Biochemistry*, *18* (1), 37-47.

Bohnsack, J.A. and Harper, O.E. (1989). Length-weight relationships of selected marine reef fishes from the southeastern United States and the Caribbean. *NOAA Technical Memorandum NMFS-SEFC*, *215* (1), 1-31.

Carpenter, R.C. (1986). Partitioning herbivory and its effects on coral reef algal communities. *Ecological Monographs*, *56* (4), 345-363.

Chellappa, S., Lima, J. T. A. X., Araújo, A. and Chellappa, N.T. (2010). Ovarian development and spawning of Serra Spanish mackerel in the coastal waters of northeastern Brazil. *Brazilian Journal of Biology, 70* (2), 451 – 456.

Cheney, K.L. and Côté, I.M. (2003). Indirect consequences of parental care: sex differences in ectoparasite burden and cleaner-seeking activity in longfin damselfish. *Marine Ecology Progress Series, 262*: 267-275.

Cleveland, A. and Montgomery, W.L. (2003). Gut characteristics and assimilation efficiencies in two species of herbivorous damselfishes (Pomacentridae: *Stegastes dorsopunicans* and *S. planifrons*). *Marine Biology*, *142*, 35-44.

Ferreira, C.E.L., Gonçalves, J.E.A., Coutinho, R. and Peret, A.C. (1998). Herbivory by the dusky damselfish *Stegastes fuscus* (Cuvier, 1830) in a tropical rocky shore: effects on the benthic community. *Journal of Experimental Marine Biological Ecology*, *229* (1), 241-264.

Fischelson, L., Popper, D. and Aviador, A. (1974). Biosociology and ecology of pomacentrid fishes around the Sinai Peninsular (northern Red Sea). *Journal of Fish Biology*, *6* (1), 119-133.

Gaillard, S., Grillasca, J.P., Leung-Tack D. and Aubert, J. (2004). A brief story of fish sex. *Journal of Fisheries Society*, *31* (1), 1-11.

Gibson, R.N., Ansell, A.D. and Robb, L. (1993). Seasonal and annual variations in abundance and species composition of fish and macrocrustacean communities on a Scottish sandy beach. *Marine Ecology Progress Series*, *98* (1), 89-105.

Goodman, L. (1965). Simultaneous confidence intervals for multinomial proportions. *Technometrics*, 7 (2), 247-54.

Hatcher, B.G. (1981). The interaction between grazing organisms and the epilithic algal community of a coral reef: A quantitative assessment. *Proceedings of the 4th International Coral Reef Symposium*, *2* (1), 515-524.

Itzkowitz, M., Ludlow, A. AND Haley, M. (2000). Territorial boundaries of the male beaugregory damselfish. *Journal of Fish Biology*, *56*, 1138-114.

Klumpp, D.W. and Polunin, N.V.C. (1989). Partitioning among grazers of food resources within damselfish territories on a coral reef. *Journal of Experimental Marine Biological Ecology*, *125*, 145- 169.

Klumpp, D.W., Mckinnon, A.D. and Daniel, P. (1987). Damselfish territories: zones of high productivity on coral reefs. *Marine Ecology Progress Series*, *40*, 41-51.

Le Cren, E.D. (1951). The length-weight relationship and population cycle in gonad weight and condition in the perch *Perca fluviatilis*. *Journal of Animal Ecology*, *20* (2), 201-219.

Lison de Loma, T., Harmelin-Vivien, M., Naim, O. and Fontaine, M.F. (2000). Algal food processing by *Stegastes nigncans,* an herbivorous damselfish: Differences between an undisturbed and a disturbed coral reef site (La Réunion, Indian Ocean). *Oceanological Acta*, *23* (7), 793-804.

Marinho-Soriano, E., Carneiro, M.A.A. and Soriano, J.P. (2009). *Manual de identificação das macroalgas marinhas do litoraldo Rio Grande do Norte.* Natal: EDUFRN, 120p.

Munro, J.J. (1976). Aspects of the biology and ecology of Caribbean reef fishes: Mullidae (goat-fishes). *Journalof Fish Biology*, *9* (1), 79-97.

Nemtzov, S.C. (1997). Intraspecifc variation in home range exclusivity by female green razorfish, *Xyrichtys splendens* (family Labridae), in different habitats. *Environmental Biology of Fish*, *50*, 371-381.

Nikolsky, G.V. (1969). *Theory of fish population dynamics*. Edinburgh, Oliver and Boyd, 180p.

Osório, R., Rosa, I.L. and Cabral, H. (2006). Territorial defense by the Brazilian damsel *Stegastes fuscus* (Teleostei: Pomacentridae). *Journal of Fish Biology*, *69*, 233-242.

Petersen, C.W. (1995). Male mating success and female choice in permanently territorial damselfishes. *Bulletin of Marine Science*, *57*, 690-704.

Picciulin, M., Varginella, L., Spoto, M. and Ferrero, E.A. (2004). Colonial nesting and the importance of the brood size in male parasitic reproduction of the Mediterranean damselfish *Chromis chromis* (Pisces: Pomacentridae). *Environmental Biololgy of Fish*, *70*, 23-30.

Reviers, B. (2006). *Biologia e filogenia das algas*. Porto Alegre, Artmed. 280p.

Richardson, D.L., Harrison, P.L. AND Harriott, V.J. (1997). Timing of spawning and fecundity of a tropical and subtropical anemonefish (Pomacentridae: Amphiprion) on a high latitude reef on the east coast of Australia. *Marine Ecology Progress Series*, *156*, 175-181.

Sale, P.F. (1977). Maintenance of high diversity in coral reef fish communities. *American Naturalist*, *111*, 337-359.

Schwamborn, S.H.L. and Ferreira, S.P. (2002). Age structure and growth of the dusky damselfish, *Stegastes fuscus,* from Tamandaré reefs, Pemambuco, Brazil. *Environmental Biology of Fish*, *63,* (1), 79-88.

Sikkel, P.C., Herzlieb, S.E. AND Kramer, D.L. (2005). Compensatory cleaner-seeking behaviour following spawning in female yellowtail damselfish. *Marine Ecology Progress Series*, *296*, 1-11.

Tabata, K. (1995). Reduction of female proportion in lower growing fish separated from normal and feminized seedlings of hirame *Paralichthys olivaceus. Fisheries Science*, *61* (2), 199-201.

Tzioumis, V. and Kingsford, MJ. (1995). Periodicity of spawning of two temperate damselfishes: *Parma microlepis* and *Chromis dispilus*. *Bulletin of Marine Science*, *57* (3), 596-609.

Volpato, G.L. and Trajano, E. (2006). Biological rhythms. In: Val LA, Val VMFA, Randal DJ (Eds) *Fish physiology*. Elsevier: San Diego, 101-153.

Werner, D. (1977). *The biology of diatoms*. Blackwell, London. 498p.

Wootton, R.J. (1995). Constraints in the evolution of fish life histories. *Netherland Journal of Zoology*, *42* (1), 291-303.

Wootton, R.J., Evans, G.W. and Mills, L.A. (1978). Annual cycle in female three-spined sticklebacks (*Gasterosteus aculeatus* L.) from an upland and lowland population. *Journal of Fish Biology*, *12* (1), 331-343.

Reviewed by Professor Dr. José Zanon de Oliveira Passavante.

Department of Oceanography, Universidade Federal de Pernambuco, UFPE, Recife, CEP: 50670-901, PE, Brazil.

SECTION IV: IMPACTS OF FISH CULTURE

In: Biology of Semiarid Tropical Fish
Editors: J. R. Gomes and S. Chellappa

ISBN: 978-1-63117-067-6

Chapter 9

IMPACT OF NILE TILAPIA CAGE CULTURE ON THE PHYTOPLANKTON COMMUNITY IN A TROPICAL RESERVOIR

Emilly Kataline Rodrigues Pessoa*[1]*, Rafson Varela dos Santos*[2]*,
***Sathyabama Chellappa*[1*] *and Naithirithi T. Chellappa*[3]**
[1]Post-Graduate Programme in Psychobiology, Centro de Biociências, UFRN, Campus Universitário, Natal, Brazil
[2]Empresa de Pesquisa Agropecuária do Rio Grande do Norte, EMPARN, Avenida Eliza Branco Pereira dos Santos, Parque das Nações, Parnamirim, RN, Brazil
[3]Department of Oceanography and Limnology, Centro de Biociências, Universidade Federal do Rio Grande do Norte, Praia de Mãe Luiza, s/n, Natal, RN, Brazil

ABSTRACT

Tropical reservoir ecosystems are important for multiple purposes, especially for public potable water supply, agriculture and fish culture. This study investigated the physical-chemical characteristics and the spatial and temporal distribution of phytoplankton community to check alterations in water quality due to impacts caused by cage culture of the exotic fish Nile tilapia, *Oreochromis niloticus* in a reservoir of Northeastern Brazil. The physical-chemical parameters (pH, temperature, electrical conductivity, concentration of dissolved oxygen and light transparency), inorganic nutrients (nitrate, ammonium ions, soluble reactive ortophosphate and total nitrogen and phosphorus) and chlorophyll *a* biomass were measured in two sites of cage culture (point source and non-point source) during two seasons (dry and rainy). Phytoplankton community showed alternating dominance among the species of Cyanobacteria, Bacillariophyceae and Chlorophyceae between the two sites. There was a considerable variation in the levels of inorganic nutrients, such as, nitrate-nitrogen, ammoniacal nitrogen and orthophosphate during the two seasons. The trophic state index diagnosed mesotrophic category, based on the values of chlorophyll *a*, total phosphorus and light

[*] E-mail of the corresponding author: chellappa.sathyabama63@gmail.com; ntchellappa@ufrnet.br.

transparency. The wind driven turbulence of the water column and the inflow of water (flushing and dilution) during the rainy season acted as constraint which did not permit the luxuriant growth of cyanobacterial species. Cage culture of tilapia did not cause pollution problems which could interfere with the reservoir water quality and the cyanobacterial bloom formation. The muted impact was due to the small scale sustained aquaculture in the well managed reservoir.

Keywords: Nile tilapia, *Oreochromis niloticus*, cage culture impact, phytoplankton community, Northeastern Brazil

INTRODUCTION

The rapid increase in population in tropical regions, and the continued demand for fish, generated the need for inland fish culture as an alternative means of increasing fish production. Nile tilapia, *Oreochromis niloticus* is popularly used for such practices in Brazil. Fish cage culture practices have diverse advantages since they use existing water bodies (reservoirs, rivers and lagoons), low cost economy involved in implementation and the use of simple technology to produce high quality protein and include extension programs of rural development (Chellappa et al., 1995; Chellappa et al., 2011). Another view held explains that this artisan activity helps to clean eutrophicated waters as it involves the culture of planktivorous tilapia species but at the same time an intense practice can cause a series of alterations to the biotic and abiotic components of the environment (Beveridge, 1987).

The universal acceptance is that water quality in lakes and reservoirs is the result of a subtle interaction between physical, chemical, and biological processes acting in spatial and temporal scales. The annual changes in phytoplankton composition and its relative abundance in freshwater aquatic ecosystems, vary from predictable to unpredictable and to chaotic distribution. The mediating factors based on the results of variation in the physical (light and temperature regime) and the chemical (nutrient availability) and these singly or cumulatively stimulate or act as constraints to the phytoplankton growth (Reynolds, 1984; Margalef, 1997; Huisman et al., 2006).

Phytoplankton functional structure and chlorophyll biomass production are often recognized as annual periodicity in species succession or assemblages (Reynolds et al., 2002; Padisák et al., 2003). Light regime experienced by tropical reservoir phytoplankton cells is differentiated into stable period (dry season) and turbulent mixing period (rainy season), which determines the residence time of phytoplankton within the well-illuminated upper layers of the water column (MacIntyre, 1998; Câmara et al., 2009). On the other hand, nutrient distributions and their bioavailability in the upper layers, where light levels are adequate, stimulate better phytoplankton growth.

This is the result of transport processes interacting with biological phenomena. In eutrophic to hypertrophic systems, where nutrients are seldom (if ever) limiting, the zmix/zeu ratio is the main constraint for phytoplankton growth controlling the phytoplankton morphological characteristics and the annual course of the phytoplankton community composition (Naselli-Flores, 2000; Naselli-Flores and Barone, 2007).

Phytoplankton seasonal variation in freshwaters has been successfully applied as functional groups to explain the changes in phytoplankton communities and this principle was based on Reynolds′ model (Reynolds, 1997). It had been related to the response of varying environmental conditions and observed in different freshwater ecosystems viz. temperate lakes (Lindenschmidt and Chorus, 1998), temperate rivers (Descy, 1993), subtropical (Hambright and Zohary, 2000) and tropical ecosystems (Figueredo and Giani, 2001). In eutrophic temperate lakes, for example, the dominant phytoplankton functional group identified may change at seasonal scales from group B diatoms to group G chlorophytes, then to group H cyanobacteria and finally to group L dinoflagellates. Phytoplankton communities can also experience changes at sub seasonal time scales, which are driven by short-term changes in the environmental conditions that interrupt, re-set, or alter the phytoplankton community dynamics. Gaedeke and Sommer (1986) demonstrated experimentally that phytoplankton diversity decreased significantly after short-term events of higher turbulence in water, resuspension of sediment materials, strong dilution that led to levels and dissolved nutrients availability. Escot et al. (2004) described the effects of strong wind events on the phytoplankton communities of a Mediterranean reservoir.

Another aspect of phytoplankton ecology was based on anthropogenic effects and the changes of phytoplankton assemblages in favor of bloom formation of toxic cyanobacterial species and considered as negative impact. Recently, Borges et al (2010) investigated the impact of cage culture of tilapia in Rosana Reservoir of the lower stretch of the Paranapanema River in Southern Brazil and observed the dominance of Cyanobacteria. Dominance of *Microcystis*, *Radiocystis* and *Pseudanabaena* has been attributed to the increased predation pressure of zooplankton on other phytoplankton species.

Very few studies have analyzed in detail the impacts caused by fish cage culture practices on the structure and chlorophyll biomass of phytoplankton community. Fish cage culture practice is a key-driver of phytoplankton dynamics in reservoirs to trigger dominant and co-dominant species of cyanobacterial species with consequent fish mortality in the semiarid reservoirs of Northeastern Brazil (Chellappa and Costa, 2003; Chellappa and Chellappa, 2011). The input of organic material, nutrients and metabolites in cage aquaculture environments promote temporary changes in the diet of natural fish populations, especially the opportunists (Strictar-Pereira et al., 2010). It was therefore felt imperative to investigate the changes on phytoplankton community structure and function under the influence of cage culture, an anthropic interference in reservoir environments.

The present study is aimed at two important aspects: firstly, to verify the physical and chemical characteristics of two sites (the point source and non-point source impact of nutrient-energy habitat template of cage culture) and the spatial and temporal variation of phytoplankton community. Secondly, to verify the relative abundance of cyanobacterial species (in the point source and non-point source), besides investigating the tropic status in relation to the effluents of cage culture.

Materials and Methods

Study Area

The present study was carried out in the Reservoir Ministro João Alves, a freshwater body where fish cage-culture is practiced by local fishermen. This reservoir is situated at national grid ref NS latitude 06° 41' 56" and 06° 41' 71" S and longitude of 36° 37' 76" and 36° 370' 84" W in the city of Parelhas of Northeastern Brazil. The catchment area has distinctive scrub vegetation, consisting of xerophytic low thorny bushes, dominated essentially by *Acasia* and *Mimosa* spp. adapted to the semi-arid climate. This reservoir was constructed in 1984 on the River Seridó of the Piranhas–Assu hydrographic basin, has a water storage capacity of 10 x 10^6 m^3 and a theoretical retention time of 780 days (Chellappa et al., 2006; Chellappa et al., 2009a; Chellappa et al., 2009b).

Table 1. Fish feeding regime used during the cage culture

Weight of fish (g)	Feed rate in relation to biomass (%)	Number of feeds per day
1 to 30	10	8
30 to 100	5	6
100 to 230	4	4
230 to 320	3	3
320 to 420	2	3
Above 420	1	2

The water outflow is maintained regularly according to the storage capacity and irrigation demands of the region. This area is dominated by humid semiarid climate with high evaporation rate and an irregular pattern of rainfall.

Cage culture of Nile tilapia, *Oreochromis niloticus* is practiced by local fishermen close to the littoral region of the reservoir. The six cages used in the fish culture were made of galvanized iron and wire mesh coated with PVC, plastic drums were used as floats and PVC plastic pipes of 100 mm were sealed at the edges. Each cage measured 2.0 x 2.0 x 1.2 m, with a capacity of 4 m^3. Each cage was installed in the reservoir at a distance of 2 meters from each other and cages were stocked with fingerlings of male Nile tilapia, *O. niloticus*, each weighing 1.0 g at a stocking density of 150/m^3, totaling to 600 fish fingerlings per cage. The cultivation period was divided into the pre-fattening and the fattening phase. The pre-fattening phase lasted for two months, during which time the fingerlings were offered extruded feed with 42% crude protein (CP). During the fattening phase which lasted for four months, the fish were offered extruded feed with 32% CP. Table 1 shows the feeding regime used during the cage culture.

Sample Collection and Analyses

Samplings were made fortnightly in two sites: site 1 (the point source) in the vicinity of the cages, and site 2 (the non-point source) which was about 0.5 km from the cages down stream. Samples were collected from the two sites, between 9.00 to 10.00 hrs at fortnightly intervals, from July 2008 to June 2009. Water temperature, pH, dissolved oxygen and electrical conductivity were measured in the field using multiprobe Kit (WTW 340 I–MERCK). Light transparency was assessed using 30 cm Secchi disc. The values of euphotic zone corresponding coefficient of light attenuation were measured by multiplying the Secchi disk depth by 2.7 (Margalef, 1983). Rainfall data was obtained from the Agricultural Research Company of Rio Grande do Norte, Brazil (EMPARN).

Water samples for inorganic nutrients such as, nitrate-nitrogen, ammoniacal-nitrogen and orthophosphate were taken with 2 dm^3 Van Dorn bottle and filtered through GF/C fiberglass filters before analyses (Goltermann et al., 1978).

Phytoplankton samples were fixed in Lugol´s iodine solution and the cell counts were done with the help of inverted microscope after 24 hours of sedimentation (Lund et al., 1958). The biovolume procedure based on simple geometric solids was used assuming unit specific gravity (Rott, 1981). The adopted taxonomic classification is based on Hoek et al., (1995). For species identification the manual of Wehr and Sheath (2003) was used. Determination of ecological indices was based on the software programme of Seaby and Henderson (2006). Water samples for chlorophyll *a* analysis were stored cold at 4^0C in the dark and filtered on Whatman GF/C filters. Concentrations were detected after overnight extraction in 90% acetone and corrected from the values of phaeophytin degradation pigment (Marker et al, 1980).

Statistical analyses were performed through Statistics 6.0 software package. Data analysis was based on log-transformed data and subjected to two-way analysis of variance (ANOVA) to test for significance between the dependent (relative abundance of phytoplankton species and independent (environmental parameters) variables.

RESULTS

The depth of the reservoir varied dramatically between the rainy and dry seasons during the study period. There was no rainfall during the dry season (July to December) and during the rainy season (January to June) a maximum rainfall of 273 mm was registered. The pattern of rainfall of the region over a period of six years was considered as a trend-setter of the seasons (Figure 1a). It is evident that the rainy season extends from January to June, with shifts in the peak of wet season from year to year. Generally the period of July to December represents the dry season in the semiarid region of Brazil. The mean monthly variation in rainfall during the study period is shown in Figure 1b. Seasonal hydrology of the study reservoir during the period of 2008 through 2009 oscillated according to the dry/wet annual cycle. The rainfall touched a peak of 294 mm during the rainy season and was low as 16 mm in the dry season (Figure 1b). This in turn, largely affected the theoretical renewal time during the period of 2008-2009 which consequently influenced the phytoplankton community.

All environmental variables in sites 1 and 2 during the rainy and dry seasons are shown in Figures 2 and 3. The trend of pH was alkaline both in spatial and temporal scales with a narrow range from 8.3 to 8.6 (Figure 2a). Transparency of water varied from 0.35 to 1.15 cm (Figure 2b) and the temperature varied from 27.2 to 30^0C (Figure 2c). During the dry season when water was in a clear state, the mean value of the concentration of dissolved oxygen varied from 6.7 mg L^{-1} in site 1 to a mean value of 7.3 mg L^{-1} in site 2. However, this situation was altered during the rainy season, when site 1 reached a maximum of 8.6 mg L^{-1} and site 2 reached a maximum of 8.0 mg L^{-1}. This coincided with large river influx into the reservoir when the water reached low temperatures (Figure 2d). Oxygen saturation calculated from the temperature and concentration of dissolved oxygen, indicated 80-120% saturation throughout the water column and depletion in oxygen saturation to 65% during the rainy season. The mean electrical conductance ranged from 671–743 μScm^{-1} during the study period (Figure

2e). These values showed no significant variation during the year, thus maintaining a constancy of 600-700 μScm^{-1} independent of evaporation during the dry season and from greater allochthonous input from the adjacent catchment areas during the wet season. Mixing events were frequent during these two seasons of sampling due to wind action and water turbulence.

The values of nitrate-nitrogen (NO_3-N) were generally low in comparison to orthophosphate (PO_4-P) of the reservoir during the study period. They ranged from 360 in site 0.1 and 0.355 mgL^{-1} in site 2 during dry season to 0.427 to 0.321 mgL^{-1} in site 1 and 2 respectively during the rainy season. Ammonium ion concentrations (NH_4-N) were less than nitrate in both sites during the dry and wet seasons, probably due to more oxidation process than reduction. Orthophosphate concentrations varied between 0.570 mgL^{-1} in site 1 and 0.546 mgL^{-1} in site 2 during the dry season to 0.320 in site 1 and 0.409 mgL^{-1} in site 2 during rainy season indicating phosphorus turn over from sediment resuspension (Figure 3).

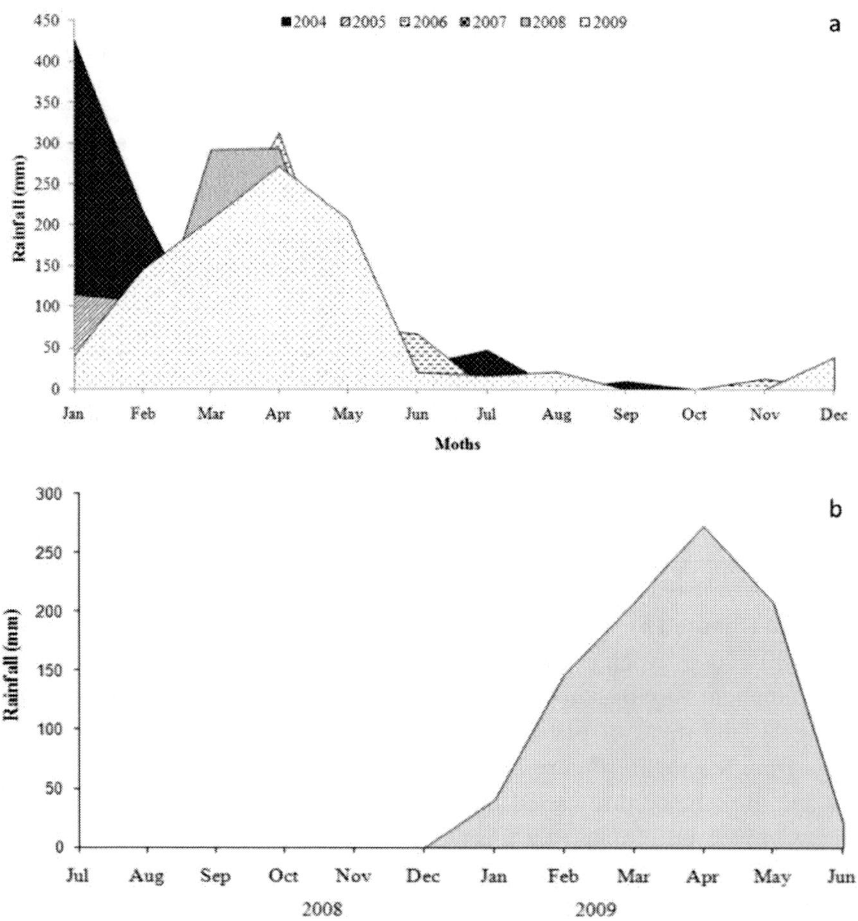

Figure 1. (a) Rainfall over a period of six years (2004 – 2009); (b) monthly rainfall during the study period.

Chlorophyll *a* concentrations ranged from 24.0 in site 1 to 11.2 μg L^{-1} in site 2 during the dry season. The levels of chlorophyll *a* in wet season increased to 29.0 in site 1 and declined to as low as 5.0 μg L^{-1} in site 2 due to turbid water condition and low light penetration (Figure 4).

The clear water phase during the dry months and the larger coefficient of light penetration stimulated more chlorophyll *a* levels in site 2 of the pelagic zone thus matching with turbid water condition of wet months.

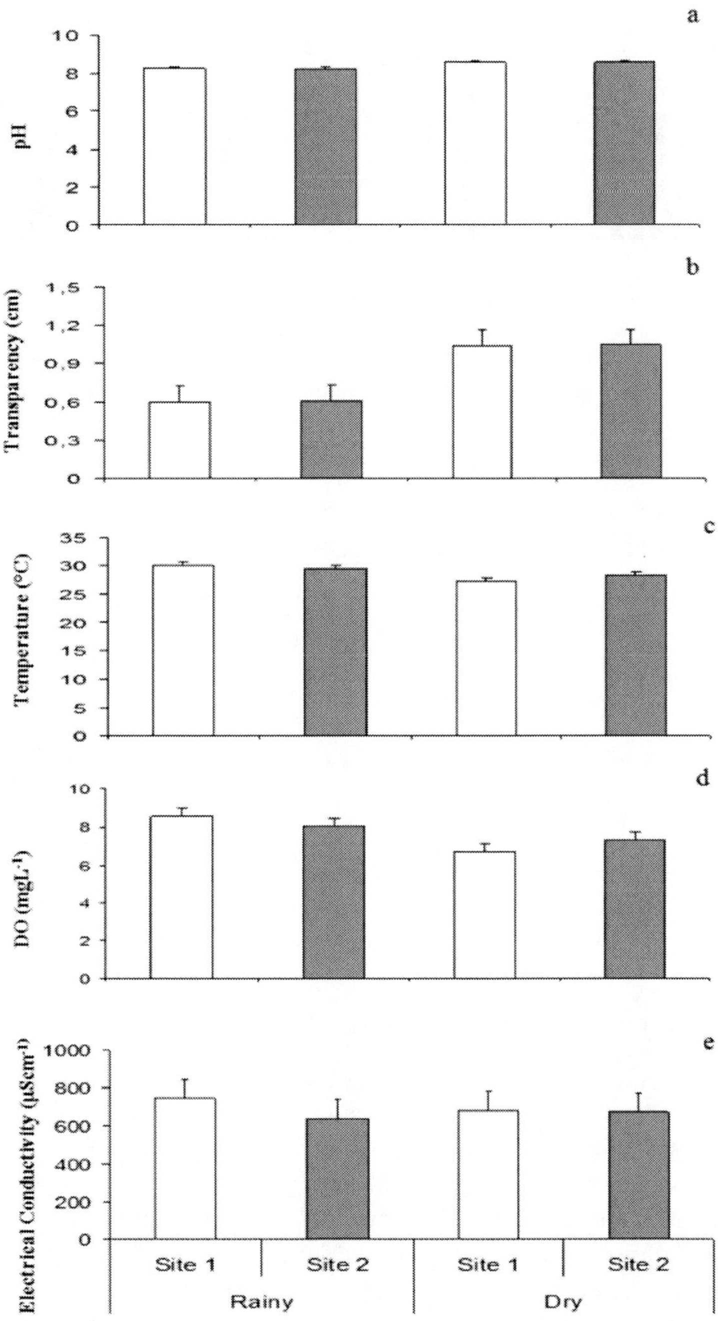

Figure 2. Values of physical-chemical parameters in sites 1 and 2 during the rainy and dry seasons: a) pH, b) transparency, c) temperature, d) concentration of dissolved oxygen, e) electrical conductivity.

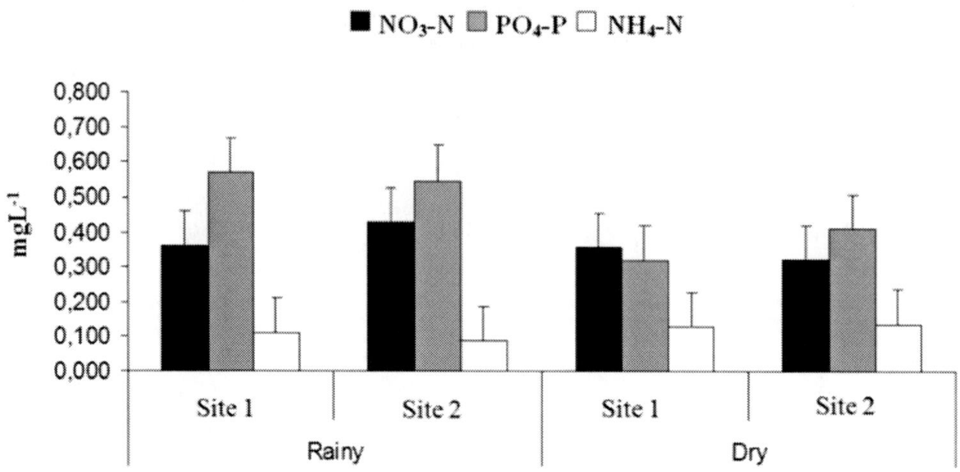

Figure 3. Mean values of inorganic nutrient variations in site 1 and site 2 during the rainy and dry seasons.

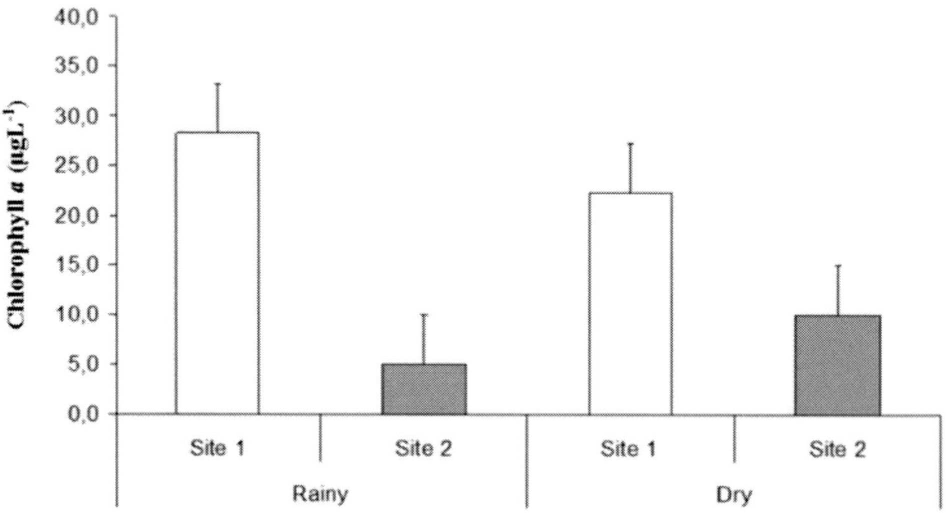

Figure 4. Mean values of chlorophyll a and transparency in site 1 and site 2 during the rainy and dry seasons.

Phytoplankton community of the reservoir comprised a total of 42 taxa and this list together with their spatial and temporal distribution is presented in Table 2. A sequence of the abundance of species follows the order of Bacillariophyceae, Chlorophyceae, Cyanobacteria, Dinophyceae, Euglenophyceae and Cryptophyceae for pelagic zone of the reservoir, representing site 2 of non-point source nutrient concentrations. This region was dominated overwhelmingly by species belonging to Bacillariophyeae such as *Aulacoseira granulata*, *Cyclotella meneghiniana* and the chlorophycean alga, *Oocystis lacustris*. This sequence had drastically changed in site 1, where the cyanobacterial species such as, *Anabaena circinalis* with coiled filaments and *A. planktonica* with straight filaments, mucilage colonies of *Microcystis aeruginosa* and very limited *Cylindrospermopsis raciborskii* individuals were recorded. This was due to the impact from cage-culture of tilapia and a point source of nutrients flux in the site 1 of littoral region. In addition to this, species numbers increased slightly from 17 to 20 taxa. The Dinophyceae was represented principally by species of *Peridinium* and *Gymnodinium* and were numerically reduced.

About 60% of the organisms were with bio volume of less than 1000 μm^3, even though the dominant species generally showed larger dimensions. The species dominance index expressed high values with filamentous cyanobacterial species in site 1 (impact of cage culture) and two species of bacillariophyceae and one of chlorophyceae in site 2 (Figure 5). Species richness index was used to express the phytoplankton diversity and it showed higher values in site 2 during wet period and synchronized with the appearance of large numbers of chlorococcales species, *Oocystis lacustris*. Species richness index was reduced considerably in site 1 both in wet and dry period due to the expressive dominance of cyanobacterial species (Figure 5).

Table 2. Phytoplankton species registered with their spatial and temporal distribution in the Reservoir Ministro João Alves, Brazil

	Rainy Period		Dry Period	
	Site 1	Site 2	Site 1	Site 2
Bacillariophyceae				
Aulacoseira granulata (Ehrenberg) Simonsen	x	x	x	x
Aulacoseira distans (Ehrenberg) Simonsen	x		x	
Cyclotella meneghiniana (Kutzing)	x	x	x	x
Cyclotella stelligera Cleve (Cleve e Grunov)	x			
Frustulia rhomboides (Ehremberg) De Toni	x	x	x	x
Navicula bacillum Ehrenberg	x	x	x	x
Navicula crytocephata Kutz	x			
Navículla viridula (Kutz) Ehrenberg	x			
Nitzschia linearis Smith	x	x	x	x
Surirella capronii Brébisson	x	x	x	x
Synedra sp	x			
Chlorophyceae				
Ankistrodesmus fusiformis Corda	x	x	x	x
Botryococcus Braunii Kutzing	x	x	x	x
*Botryococcus protuberans*West and West	x	x		
Chlorella vulgaris Beijerinck	x		x	
Closterium parvulum Näg.	x	x	x	x
Coelastrum sp	x		x	
Dictyosphaerium pulchellum Naegli	x			
Oocysti borgei Snow	x	x	x	x
*Oocysti lacustri*s Chodat	x	x		
Pandorina morum Bory	x		x	
Pediastrum duplex Meyen	x		x	
Radiococcus sp Midle	x		x	

Table 2. (Continued)

	Rainy Period		Dry Period	
	Site 1	Site 2	Site 1	Site 2
Scenedesmus acunts Meyen	x	x	x	x
Scenedesmus dimorphus (Turp.) Kutz	x	x	x	
Scenedesmus quadricauda (Tupin) Brébisson	x			
Staurastrum leptocladum Nordst.	x		x	
Staurodesmus triangulares (Smith) Tham	x			
Cyanophyceae				
Anabaena circinalis (kütz) Ralb	x	x	x	x
Anabaena planktonica Lemm	x	x		x
Cylindrospermopsis raciborskii Horecká	x		x	
Chorooccocus turgidus (Kutz.) Nag	x		x	
Microcystis sp	x	x	x	x
Oscillatoria articulata Gardner	x		x	x
Oscillatoria lacustris (Kleber.) Geitler	x		x	
Oscillatoria limnetica Lemmerman	x	x		
Pseudoanabaena catenata Laut.	x	x	x	x
Dinophyceae				
Gonyaulax sp	x			
Gymnodinium sp	x		x	
Peridinium volzii Lemmermann	x		x	
Euglenophyceae				
Euglena acus Ehrenberg	x		x	
Phacus tortus (Lemmermann) Skvortzov	x	x		x

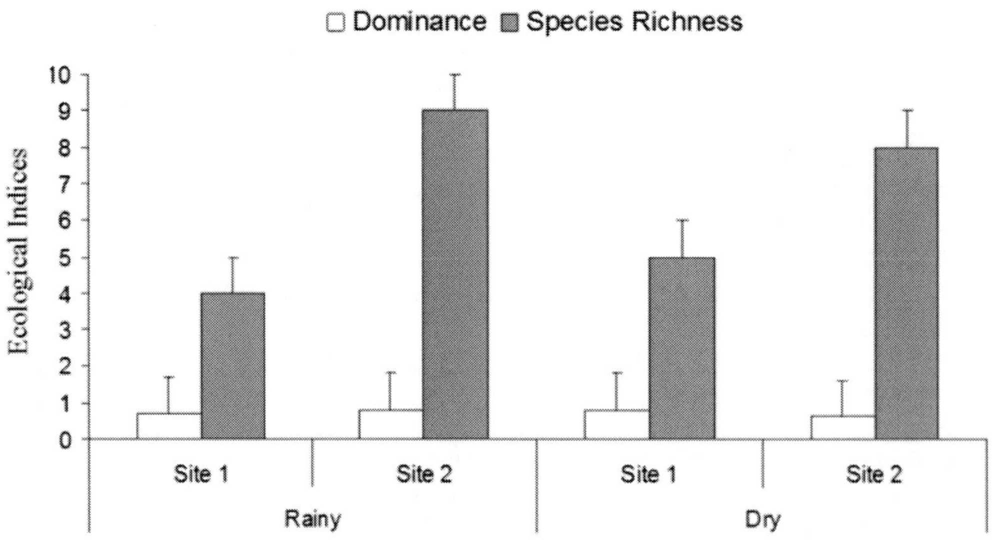

Figure 5. Ecological indices of phytoplankton during the rainy and dry seasons.

Table 3. Significant F-values ($p < 0.05$) from one-way Anova and Kruskal-Wallis test for spatio-temporal variations of physical-chemical characteristics and chlorophyll *a* concentrations in the Reservoir Ministro João Alves, Brazil

	Rainy period					Dry period				
	DF*	SS*	MS*	F	p	DF	SS	MS	F	p
pH	1	0.0015	0.0015	0.003	0.957	1	0.0008	0.0008	0.0199	0.89
Temperature	1	0.72	0.72	0.326	0.589	1	2.205	2.205	0.983	0.36
Dissolved oxygen	1	0.50000	0.5	0.943	0.369	1	0.72	0.72	0.735	0.42
Electrical cond.	1	41905.1	41905.1	1.7	0.24	1	136.125	136.125	0.211	0.66
Transparency	1	0.0002	0.0002	0.006	0.94	1	0.00005	0.00005	0.0024	0.96
Nitrate	1	0.00905	0.00905	0.278	0.617	1			*0.337	0.69
Orthophosphate	1	0.00109	0.00109	0.071	0.798	1	0.0168	0.0168	2.365	0.18
Ammonia	1			*1.33	0.343	1	0.00010	0.00010	0.0202	0.89
Chlorophylla *a*	1	1081.13	1081.125	59.65	<0.001	1	1038.2	1038.22	36.287	0.04

*Kruskal-Wallis.

GL: degrees of freedom, SS: statistically significant, MS: median significant.

Table 4. Significant t-test values ($p<0.05$) for spatio-temporal variations of physical-chemical characteristics and chlorophyll *a* concentrations

	Rainy period				Dry period			
	Mean P1 (±Std)	Mean P2 (±DP)	t-value	p-value	Mean P1(±Std)	Mean P2 (±DP)	t-value	p-value
pH	8.27 (±0.62)	8.24 (±0.75)	0.06	0.957	8.56 (±0.24)	8.54 (±0.15)	0.14091	0.892554
Temp.	30 (±1.51)	29.4 (±1.45)	0.57	0.589	27.22 (±0.86)	28.275 (±1.93)	-0.99160	0.359681
D.O	8.55 (±0.50)	8.05 (±0.89)	0.97	0.369	6.7 (±1.32)	7.3 (±0.46)	-0.85714	0.424273
Elec. Cond.	743 (± 44.1)	637.75 (±21.5)	0.43	0.684	678.75 (±27.01)	670.5 (±23.6)	0.45989	0.661790
Transp.	0.59 (±0.17)	0.60 (±0.18)	-0.08	0.940	1.03 (±0.16)	1.04 (±0.12)	-0.04904	0.962480
Nitrate	0.35 (±0.18)	0.42 (±0.18)	-0.53	0.617	0.35 (±0.13)	0.32 (±0.15)	0.34733	0.740204
Orthophosphate	0.56 (±0.12)	0.54 (±0.12)	0.27	0.798	0.32 (±0.5)	0.40 (±0.10)	-1.53777	0.175021
Ammonium	0.11 (±0.006)	0.08 (±0.04)	0.93	0.388	0.13 (±0.07)	0.13 (±0.06)	-0.14218	0.891589
Chloro *a*	28.25 (±4.85)	5 (±3.55)	7.72	0.0002	22.25 (±5.25)	10 (±6.21)	3.01005	0.023699

Table 3 shows the results of one-way ANOVA, which is singularly significant in relation to chlorophyll *a* levels between sites 1 and 2. Significant t-test values ($p<0.05$) for spatio-temporal variations of physical-chemical characteristics and chlorophyll *a* concentrations are given in Table 4.

DISCUSSION

The present paper documents the impact of Nile tilapia cage culture on the successional pattern of phytoplankton species composition on spatial scale in the shallow tropical reservoir of Northeastern Brazil. This finding of potential effect of tilapia cage culture highlights the discussion on three broad contexts: Firstly, the inorganic nutrient levels in site 1, where the point source nutrient is significant when compared to the site 2 of the non-point nutrient source. Secondly, the effect of the low N:P ratio on the dominant cyanobacterial species is robust and persistent. Finally, the influence of flushing of reservoir during rainy season and the subsequent shift in phytoplankton species composition on temporal scale is significant.

The relative dominance of phytoplankton varied exceptionally from cyanobacterial species dominance in the littoral region (site1) to diatoms dominated pelagic zone (site 2), which was a different situation from the one observed in an earlier study, where the dominance of diatom population was overwhelming in all zones (Nascimento, 2003; Chellappa et al., 2006). However, the present results are similar to the observation of Borges et al. (2010), where the dominance of cyanobacteria emanated from fish cage-culture practice.

International experience on Lake Aquaculture indicates that the cage culture is well established in Chile, despite environmental problems due to nutrient enrichment and organic waste impacts (EAO, 1997).

In southern Australia, particularly in Tasmanian lakes, diverse trophic status had been reported. The effects of cage culture resulted either in oligotrophic status with relatively low algal biomass or remained as mesotrophic (e.g. Lakes Sorell and Crescent), well mixed and with abundant phytoplankton.

Few cases have been reported as eutrophic at some stage (e.g. Craigbourne Dam). The eutrophic lakes have been associated with algal blooms (both blue-green and green) for varying periods, which are highly susceptible to changes in nutrient status, chlorophyll *a* levels and TP concentrations (Davies, 2000; Sanger, 1992). The present paper when compared to these studies, indicates that low N:P ratio at point source site exhibited selective eutrophication during dry period, and consequently non-toxic cyanbacterial species appeared without resulting in fish mortality.

Fish kills have been registered in other reservoirs of Northeastern Brazil (Chellappa et al., 2000; Chellappa and Costa, 2003; Chellappa and Chellappa, 2011). The situation in the non-point source of the pelagic region was totally opposite, where species of diatoms and chlorophyceae dominated and the environment was free from the impact of cage culture effluents.

Another ecological explanation offered is the regime shift of temporal changes during the time when influx of river flow to reservoir increases considerably and stimulates phytoplankton species richness. This explanation considers that abrupt and persistent shift could be induced through eutrophication related drivers (Jeppesen et al., 1998; Crossetti and Bicudo, 2008).

The shift may be from prokaryotic filamentous cyanobacterial dominance to unicellular and colonial eukaryotic phytoplankton species (Scheffer et al., 1997). However, in the present study such changes were observed as an impact from cage-culture effluent, which is selective and significant only at point source site with the persistent dominance of *Anabaena circinalis* and *A. planktonica*. The reduction of species richness is partially in agreement with the results from semiarid Mediterranean phytoplankton species composition with cyanobacterial dominance (Barone et al., 2010).

Phytoplankton biomass is represented by the chlorophyll *a* levels and is always linked to the light limitation factor in northern hemisphere reservoirs. Whereas, in semiarid tropical climate, light is year around abundant and dissolved inorganic nutrients levels of nitrogen and phosphorus act as strong environmental drivers in increasing chlorophyll *a* concentrations.

The present study clearly indicates that increased nutrients especially orthophosphate in the vicinity of fish cages enhanced levels of chlorophyll *a* in point source site than in non-point source site of the pelagic region.

As a response to two opposing resource-gradients, a situation similar to the high values registered in Scottish Lochs (SEPA, 1997) was observed in this study. Significant resource-gradient differentiation in spatial as well as temporal distribution of chlorophyll *a* was registered in the point source site and non-point source ($P < 0.05$).

This study deals with comparative analysis of phytoplankton species composition based on spatial and temporal scale for an annual cycle in a tropical reservoir, wherein Nile tilapia cage culture was practiced by local fishermen. The effluents which originated from the cage culture of tilapia showed increased levels of nitrogen and phosphorus with more of low N:P ratio. Site 1 presents predominant growth of filamentous cyanobacterial species with no harmful effects to fish but resulting in reduced phytoplankton diversity. The site 2 revealed the dominance of diatoms, presence of some fast growing species of chlorophyceae and improved species diversity.

The levels of chlorophyll *a* are higher in the site 1 than in site 2. These spatial variations in structural and functional components of phytoplankton and are attributed to opposing resource gradients of site 1 and site 2. This is an important social inclusion project sustained by the local municipality to promote sustained development project of cage-culture to improve the standards of fishermen community of the region.

The report from the present study might be considered as useful proposal of a best practice environmental management approach to reservoir fish farming through moderate cage culture, with an emphasis on methods to reduce potential risks (toxic cyanobacterial) and impacts (related to eutrophication).

ACKNOWLEDGMENTS

This study was supported by the National Council for Scientific and Technological Development of Brazil (CNPq) in the form of Research grants (N.T. Chellappa, S. Chellappa and R. V. Santos) and by the Foundation CAPES/MEC (scholarship awarded to E. K. R. Pessoa during the study period).

REFERENCES

Barone, R., Castelli, G. and Naselli-Flores, L. (2010). Red sky at night cyanobacteria delight: the role of climate in structuring phytoplankton assemblage in a shallow, Mediterranean lake (Biviere di Gela, southeastern Sicily). *Hydrobiologia*, 639, 43-53.

Beveridge, M. C. M. (1987). Cage aquaculture, Fishing News Book, Cambridge.

Borges, P.A.F., Train, S., Dias, J.D. and Bonecker, C.C. (2010). Effects of fish farming on plankton structure in a Brazilian tropical reservoir. *Hydrobiologia*, 649, 279-291.

Câmara, F.R.A., Lima, A.K.A., Rocha, O. and Chellappa, N.T. (2009). The role of nutrient dynamics on the phytoplankton biomass (chlorophyll-*a*) of a reservoir-channel continuum in a semi-arid tropical region. *Acta Limnologica Brasiliensia*, 21, 431-439.

Chellappa, N. T. and Chellappa, S. (2011). Bloom events of toxin producing cyanobacterial species associated with fish kills in a tropical reservoir from Brazil. In: Kaul, B.L. (Ed.), *Advances in Fish and Wildlife Ecology and Biology,* Vol. 5, 95-106, Delhi, Daya Publishing House.

Chellappa, N. T., Costa, M.A.M. (2003). Dominant and co-existing species of Cyanobacteria from a eutrophicated reservoir of Rio Grande do Norte State, Brazil. *Acta Oecologica*, 24, 3-10.

Chellappa, N. T., Costa, M.A.M., Marinho, I. R. (2000). Harmful cyanobacterial blooms from semiarid freshwater ecosystems of North-East Brazil. Australia. *Australian Society of Limnology Newsletter*. 38, 45-49.

Chellappa, N.T., Chellappa, T., Câmara, F.R.A., Rocha, O. and Chellappa, S. (2009b). Impact of stress and disturbance factors on the reservoir phytoplankton communities in Northeastern Brazil. *Limnologica*, 39, 273-282.

Chellappa, N.T., Chellappa, T., Lima, A.K.A., Borba, J.L.M., Souza, P.V.V. and Chellappa, S. (2006). Ecology of the freshwater phytoplankton assembleges from a tropical reservoir of northeastern Brazil. *International Journal of Lakes and Rivers*, 1 (1), 61-73.

Chellappa, S., Bueno, R.M.X., Chellappa, T., Chellappa, N.T. and Val, V.M.F.A. (2009a). Reproductive seasonality of the fish fauna and limnoecology of semi-arid Brazilian reservoirs. *Limnologica*, 39, 325- 329.

Chellappa, S., Chellappa, N.T., Barbosa, W.G., Huntingford, F.A. and Beveridge, M.C.M. (1995). Growth and production of the Amazonian tambaqui in fixed cages under different feeding regimes. *Aquaculture International*, 3, 11-21.

Chellappa, S., Nascimento, W.S., Chellappa, T. and Chellappa, N.T. (2011). Impacts of anthropic factors on native freshwater fish in Brazilian semiarid region. In: Dempsey, S.P. (Ed.), *Fish Ecology*, 115 – 130, New York, Nova Science Publishers.

Crossetti, L.O. and Bicudo, C.E.M. (2008). Phytoplankton as a monitoring tool in a tropical urban shallow reservoir (Garças Pond): the assemblage index application, *Hydrobiologia*, 610, 161-173.

Davies, P.E. (2000). Lake cage culture review report. 1-35. Ministry of Inland Fisheries, UK.

Descy, J. P. (1993). Ecology of the phytoplankton in the River Moselle: effects of disturbance on community structure and diversity. *Hydrobiologia*, 249, 111–116.

EAO (1997). *Salmon Aquaculture Review*. Chile. Vol. 4, Part C, Chapter VII. Environmental Assessment Office, British Columbia.

Escot, C., Risco, M., Armengol, J. and Basanta, A. (2004). El efecto meteorológico como un factor clave para desarrollo de pulsos de fitoplancton en embalses de abastecimiento. *Actas del XII Congreso de la Asociación Española de Limnología y IV Congreso Ibérico de Limnología.* 1-127, Porto, Portugal.

Figueredo, C.C. and Giani, A. (2001). Seasonal variation in the diversity and species richness of phytoplankton in a tropical eutrophic reservoir. *Hydrobiologia*, 445, 165–174.

Gaedeke, A. and Sommer, U. (1986). The influence of the frequency of periodic disturbances on the maintenance of phytoplankton diversity. *Oecologia*, 71, 25–28.

Golterman, H. L., Clymo, R.S. and Ohnstad, M.A.M. (1978). Methods for physical and chemical analysis of Freshwaters. IBP Handbook, 215. *Blackwell Sci. Publ.* Oxford.

Hambright, K.D. and Zohary, T. (2000). Phytoplankton species diversity control through competitive exclusion and physical disturbance. *Limnology and Oceanography*, 45, 110-122.

Hoek, V.D., Mann, C.D. and Jahn, H.M. (1995). *Algae: An introduction to phycology.* Cambridge University Press, London.

Huisman, J., Pham Thi, N.N., Karl, D.M. and Sommeijer, B. (2006). Reduced mixing generates oscillations and chaos in the deep chlorophyll maximum. *Nature*, 439, 322-325.

Jeppesen, E., Søndergaard, M., Søndergaard, M. and Christoffersen, K. (Eds.) (1998). *The Structuring Role of Submerged Macrophytes in Lakes.* Ecological Series, vol. 131. Springer-Verlag, 1-423.

Lindenschmidt, K.L. and Chorus, I. (1998). The effect of water column mixing on phytoplankton succession, diversity and similarity. *Journal of Plankton Reserch*, 20, 1927-1951.

Lund, J.W.G., Kipling, C. and Le Cren, E.D. (1958). The inverted microscope method of estimating algal numbers and the statistical basis of estimations by Counting, *Hydrobiologia*, 11, 143-170.

MacIntyre, S. (1998). Turbulent mixing and resource supply to phytoplankton. In: Imberger, J. (Ed.), *Physical Processes in Lakes and Oceans, Coastal and Estuarine Studies*, vol. 54. 561–590, American Geophysical Union, Washington, DC.

Margalef, R. (1983). Limnologia. *Ediciones Omega*, Barcelona, Spain.

Margalef, R. (1997). Our biosphere. In: Kinne, O. (Ed.), *Excellence in Ecology.* 1-176, Ecology Institute, Oldendorf Luke.

Marker, A.F.H., Nusch, E.A., Rai, H. and Rieman, B. (1980). The measurements of photosynthetic pigments in freshwater and standardization of methods: conclusions and recommendations. *Archiv für Hydrobiologie*, 14, 91-106.

Nascimento, H.C. (2003). *Ecological characterization of the phytoplankton community of João Alves reservoir,* Parelhas, RN. M.Sc. dissertation, Federal University of Rio Grande do Norte, Brazil.

Naselli-Flores, L. (2000). Phytoplankton assemblages in twenty one Sicilian reservoirs: relationships between species composition and environmental factors. *Hydrobiologia*, 424, 1–11.

Naselli-Flores, L. and Barone, R. (2007). Pluriannual morphological variability of phytoplankton in a highly productive Mediterranean reservoir (Lake Arancio, Southwestern Silicy). *Hydrobiologia*, 578, 87-95.

Padisák, J., Borics, G., Fehér, G., Grigorszky, I., Oldal, I., Schmidt A. and Zámbóné-Doma, Z. (2003). Dominant species, functional assemblages and frequency of equilibrium phases in late summer phytoplankton assemblages in Hungarian small shallow lakes. *Hydrobiologia*, 502, 157-168.

Reynolds, C. S. (1997). *Vegetation Processes in the Pelagic. A Model for Ecosystem Theory.* Excellence in Ecology, Oldendorf.

Reynolds, C.S. (1984). *The Ecology of Freshwater Phytoplankton.* Cambridge University Press, Cambridge.

Reynolds, C.S., Huszar, V., Kruk, C., Naselli-Flores, L. and Melo, S. (2002). Towards a functional classification of the freshwater phytoplankton. *Journal of Plankton Reserch*, 24, 417-428.

Rott, E. (1981). Some results from phytoplankton counting intercalibration. Schweiz. Zur. *Hydrobiologia*, 43, 34-62.

Sanger, A. (1992). Inland Fisheries Commission Biological Consultancy. *Annual Report to the Hydro Electric Commission.* 1-91. HEC/IFC Hobart.

Sanger, A. (1992). Inland Fisheries Commission Biological Consultancy. *Annual Report to the Hydro Electric Commission.* 1-91, HEC/IFC Hobart.

Scheffer, M., Rinaldi, A., Gragnani, A., Mur, L.R. and Van Nes, E.H. (1997). On the dominance of filamentous cyanobacteria in shallow turbid lakes. *Ecology*, 78, 272-282.

Seaby, R.M.H. and Henderson, P.A. (2006). *Species diversity and species richness measuring and understanding Biodiversity*. Pisces Conservation Ltd., UK.

SEPA, Scottish Environmental Protection Agency. (1997). Total phosphorus water quality standards for Scottish freshwater lochs, Policy 16, Dingwall, Scotland.

Strictar-Pereira, L., Agostinho, A.A. and Gomes, L.C., 2010. Cage culture with tilapia induces alteration in the diet of natural fish populations: the case of *Auchenipterus osteomystax. Brazilian Journal of Biology*, 70, 1021-1030.

Wehr, J.D. and Sheath, R.G. (2003). *Freshwater algae of North-America: Ecology and Classification*. Academic Press. London.

Reviewed by Professor Dr. José Zanon de Oliveira Passavante
Department of Oceanography, Universidade Federal de Pernambuco,
UFPE, Recife, CEP: 50670-901, PE, Brazil

SECTION V: ASPECTS OF FISH PARASITES

In: Biology of Semiarid Tropical Fish
Editors: J. R. Gomes and S. Chellappa
ISBN: 978-1-63117-067-6

Chapter 10

OCCURRENCE OF *NEOECHINORHYNCHUS CUREMAI* (ACANTHOCEPHALA: NEOECHINORHYNCHIDAE) IN *MUGIL CUREMA* (OSTEICHTHYES: MUGILIDAE) FROM THE COASTAL WATERS OF RIO GRANDE DO NORTE, BRAZIL

Elizete Teresinha Santos Cavalcanti*[1], *Ricardo Massato Takemoto*[2], *Maria de los Angeles Perez Lizama*[2], *Leucio Câmara Alves*[3], and *Sathyabama Chellappa*[4]

[1]Unidade Acadêmica Especializada em Ciências Agrárias,
Universidade Federal do Rio Grande do Norte, Brazil
[2]NUPELIA, Universidade Estadual de Maringá, UEM, NUPELIA, Maringá, PR, Brazil
[3]Programa de Pós-Graduação em Ciências Veterinárias,
Universidade Federal Rural de Pernambuco, UFRPE, Recife, PE, Brazil
[4]Departamento de Oceanografia e Limnologia, Centro de Biociências,
Universidade Federal do Rio Grande do Norte, UFRN, Rio Grande do Norte, Brazil

ABSTRACT

Global climate changes are predicted to impact disease dynamics, particularly for parasitic infections in marine environments. This paper reports on the parasitic infection by *Neoechinorhynchus curemai (*Acanthocephala: Neoechinorhynchidae*)* in the marine fish white mullet, *Mugil curema* from the coastal waters of Northeastern Brazil. This study alerts on the possibility of the zoonotic potential of this parasite since the host fish is of commercial importance and it is frequently used for human consumption. A total of 54 fish specimens were analyzed, out of which the intestines of 11 were infected by *N. curemai*. The highest mean intensity was 5 (range 1-15) and the highest prevalence was 42. Female fishes were more parasitized than males ($p < 0.05$). This paper reports on the occurrence of *N. curemai* in the coastal waters of Northeastern Brazil, which are

* E-mail address of corresponding author: chellappa.sathyabama63@gmail.com.

impacted by alterations of the marine environment resulting from climate change and alerts on the possibility of zoonotic potential of this parasite.

Keywords: *Neoechinorhynchus curemai, Mugil curema,* parasitism, coastal waters, Brazil

INTRODUCTION

The Acanthocephalans parasitize marine and freshwater fishes which occur from the Arctic to the Antarctic Oceans. They exhibit elongated cylindrical body, covered with thick cuticle and pleated across, and in the rear end posses a retractile proboscis covered with spines, which has the function of fixing the parasite to the host (Nickol, 1985). The acanthocephalans are considered to be pathogenic since they are endowed with this mechanism (Thatcher, 1991). Acanthocephalans typically have complex life cycles involving a number of hosts, including invertebrates, fish, amphibians, birds and mammals. Some species can infect humans and for this reason the study of the Acanthocephala has its importance in public health (Tantaleán et al., 2005).

Some species of Acanthocephala are responsible for producing acanthocephaliasis in humans, and those cited so far are *Moniliforme moniliformis*, *Plagiorhynchus cylindraceu*, and *Pomphorhynchus laevis*. Holloway and Bier (1967), and Schmidt and Roberts (1981), indicate that the genus *Corynosoma* is potentially pathogenic to humans. A case of human infection with an acanthocephalan was reported, when *Bolbosoma* sp. (class Palaeacanthocephala) was found at laparotomy, in an old Japanese fisherman who complained of acute abdominal pains. The parasite probably was acquired by eating raw bluefin tuna fish (Tada et al., 1983).

In natural environments, *N. curemai* was reported in *Prochilodus scrofa* (Noronha, 1984) and in *P. lineatus* (Martins et al., 2000; Lizama et al., 2005). This parasite was found in the fish white mullet, *Mugil curema* in the Potengi Estuary, Rio Grande do Norte, Brazil (Fortes et al., 2000). However, there are no records of occurrence of *N. curemai* in coastal waters and no known case of human infection has been reported (Sterrer, 1986). This study registers the occurrence of the Acanthocephalan parasite *N. curemai* in *M. curema* from the coastal waters of Northeastern Brazil.

MATERIALS AND METHODS

This work was carried out in the littoral region of Rio Grande do Norte, Northeastern Brazil (05°47'42" S and 35°12'24" W). During August to October 2011, a total of 54 specimens of white mullet, *M. curema* (Osteichthyes: Mugilidae) were captured with the help of local fishermen. Fish were captured using beach trawl, a type of trammel net with 110 m long, 10 mm mesh in the center and 70 mm at the ends. Fish collected were numbered, the total body length (in centimetres) and body mass (in grams) of fish were registered. Samples of whole fish were used for morphometric analysis to confirm the taxonomic identification of the species (Szpilman, 2000). The gonads of host fish were examined to determine the sex and maturity stages (Vazzoler, 1996; Marin et al. 2003). To determine the difference in the ratio between males and females of host fish and to verify if the difference was statistically significant, the chi-square ($\chi 2$) test was applied at 5% level of significance using the software Statistica 7.0.

The body surface, gill chambers and digestive tract of each fish was checked for the presence of parasites. The fish were necropsied and dissected out to examine the gonads and digestive tract, and when parasites were detected they were carefully removed and observed under a stereoscopic microscope, processed and quantified. Acanthocephalan parasites were placed on Petri dishes with distilled water, left in the refrigerator for 24 hours and then fixed in 5% formalin. Later, the parasites were cleared by creosote of Faia, stained with Mayer´s carmalumen and mounted in Canada Balsam (Eiras et al. 2006). Parasitic indices were calculated and expressed as per Bush et al. (1997). The identification of Acanthocephalan parasites was carried out according to Yamaguti (1961), Amin (1987) and Thatcher (1991).

RESULTS

In the 54 fish specimens examined, 20.37% (n=11) were parasitized by the Acanthocephalan *Neoechinorhynchus curemai* (Figure 1a and b). This parasite was found in the intestinal tract as well as in the serosa of the intestine, fixed by their proboscis (Figure 2).

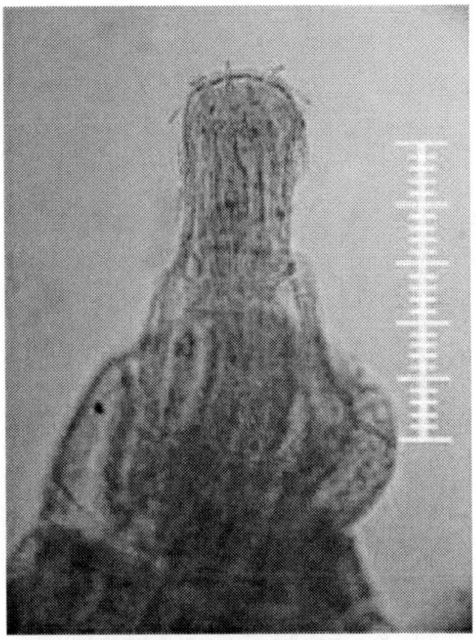

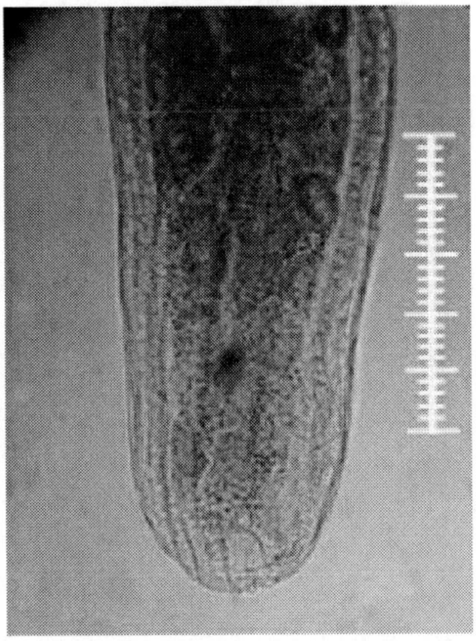

Figure 1. Neoechinorhynchus curemai (a) anterior hooks from the proboscis and (b) posterior end of the parasite. Scale bar: 0.5 mm.

Of the 11 infected fish, eight were females (72.73%), five were in the maturing stage (62.5%), two were sexually mature (25%) and one was in the spent stage (12.5%). Of the three infected males (27.27%), two were sexually mature (66.66%) and one was spent (33.33%) (Figure 3). More females of *M. curema* were parasitized than males ($\chi 2 = 79.46$; $p < 0.05$). The prevalence, mean intensity, mean abundance, number of infected males and females are shown in Table 1.

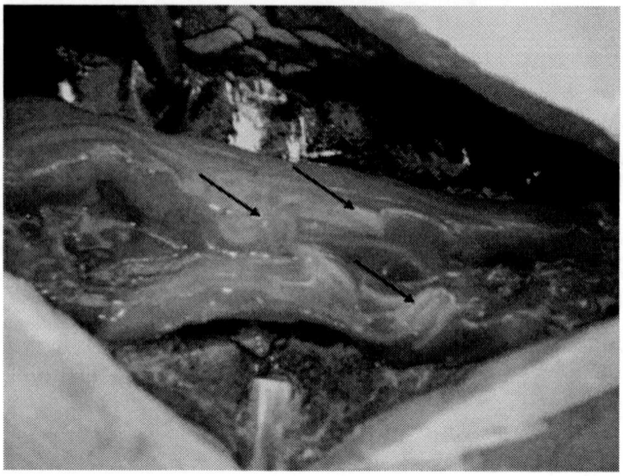

Figure 2. *Neoechinorhynchus curemai* in the serosa of the intestine of Mugil curema.

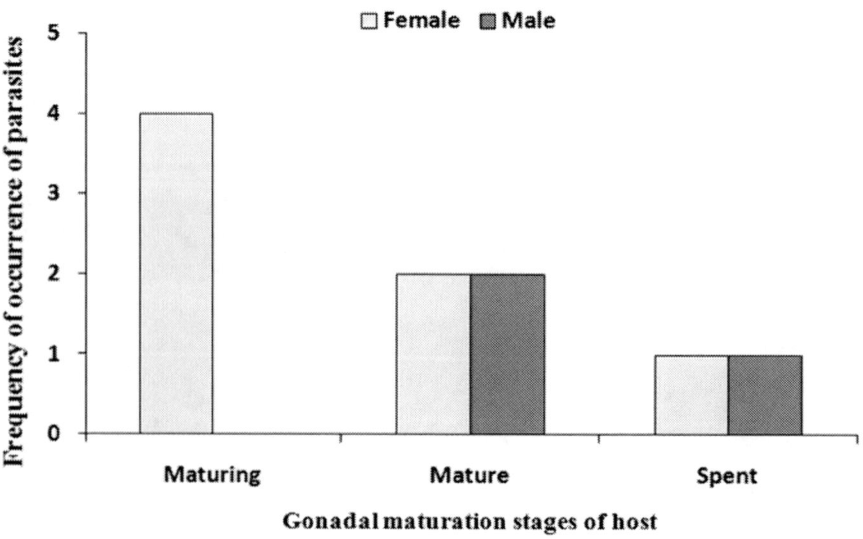

Figure 3. Frequency of occurrence of *Neoechinorhynchus curemai* in females and males in Mugil curema in different stages of gonad maturation.

Table 1. Parasitological assessment of *Mugil curema* infected by *Neoechinorhynchus curemai* in the coastal waters of Rio Grande do Norte, Brazil

Months	IF/EF (P%)	MI (range)	MA	IM/EM (P%)	IFE/EFE (P%)
August, 2011	4/20 (10)	5 (1-15)	1.00	1/8 (5)	3/12 (15)
September, 2011	2/22 (9.1)	3 (2-4)	0.27	0/11 (0)	2/11 (10)
October, 2011	5/12 (42)	3.2 (1-8)	1.33	2/7 (10)	3/5 (15)
Total	11/54 (20.37)	3.82 (1-15)	2.70	3/26 (7.5)	8/28 (13.33)

IF/EF (P%): infected fish/examined fish (P% prevalence); MI: mean intensity; MA: mean abundance; IM/EM: infected males/examined males; IFE/EFE: parasitized females/examined females.

DISCUSSION

Fishing activities of Northeastern Brazil are generally restricted to a small-scale artisanal fishery, which is conducted from small coastal villages. Most of the catch is consumed directly by the fishermen and their families, and a part is sold at local markets. Although fish is of primary importance, the local distribution of the fishery products often produces only minor interest and consequently with limited fisheries research. The growing importance of fish as a protein resource, and the interest in exporting fishery products to markets with high quality standards, requires knowledge of fish health in exploited stocks (Palm, 1997). The present study documents the occurrence of acanthocephalan parasite *N. curemai* with zoonotic potential in *M. curema* from the coastal waters of Northeastern Brazil.

Fish belonging to the family Mugilidae spawn in the sea, the young after they acquire the ability to swim actively, migrate towards the coastal waters and enter the estuaries, where they grow and the adults migrate back to the sea to spawn (Harrison, 2002). This migratory behavior of mullet explains the presence of *N. curemai* parasites in the coastal as well as in the estuarine ecosystems (Fortes et al., 2000).

The prevalence of *N. curemai* in the coastal waters of Northeastern Brazil was lower than that observed in the freshwaters of Volta Grande, Minas Gerais, Brazil (Martins et al. 2000), and River Parana in São Paulo (Santos et al., 2005). However, it was similar to that found in the Floodplains of the Upper Parana River in Brazil (Lizama et al. 2005).

In this study, younger and maturing females of *M. curema* were parasitized, and similar results were obtained in studies carried out by Amin (1985), Lasee (1989), Brasil-Sato and Pavanelli (1999), wherein a higher number of parasites were also found in female hosts. The preference of parasites for younger female individuals possibly occur since they have more energy sources, such as lipids, in contrast to mature and spent individuals. The mature and spent fish have depleted energy reserves due to the use of lipids for gonad maturation and reproduction (Chellappa et al., 1995; Huntingford et al., 2001).

The occurrence of *N. curemai* in *M. curema* in the coastal waters of northeastern Brazil, warrants the development of detailed studies of this parasite, since other Acanthocephalan species have been recorded as responsible for acanthocephaliasis in humans (Holloway and Bier, 1967; Schmidt and Roberts, 1981).

ACKNOWLEDGMENTS

This study was supported by the Post-Graduate Federal Agency CAPES/MEC, Brazil, and the National Council for Scientific and Technological Development of the Ministry of Science and Technology of Brazil (CNPq/ MCT) in the form of a Research grants to the authors.

REFERENCES

Amin, O. M., 1985: Acanthocephala from lake fishes in Wisconsin: *Neoechinorhynchus roberbaueri* n. sp. from *Erimyzon sucetta* (Lacepede), with a key to species of genus

Neoechinorhynchus Hamann, 1892 from North American freshwater fishes. *Journal of Parasitology*. 71, 312-318.

Amin, O. M., 1987: Key to the families and subfamilies of acanthocephala with the erection of a new class (Polyacanthocephala) and a new order (Polyacanthorhynchida). *Journal of Parasitology*. 73, (6), 1216-1219.

Brasil-Sato, M. C., Pavanelli, G. C., 1999: Ecological and reproductive aspects of *Neoechinorhynchus pimelodi* (Eoacanthocephala, Neoechinorhynchidae) of *Pimelodus maculatus* Lacèpède, (Siluroidei, Pimelodidae) from the basin of the São Francisco River, Brazil. *Revista Brasileira de Zoologia*. 16, 73-82.

Bush, A. O., Lafferty, K. D., Lotz, J. M., Shostak, W., 1997: Parasitology meets ecology on its own terms: Margolis et al., revisited. *Journal of Parasitology*. 83, 575-583.

Chellappa, S., Huntingford, F. A., Strang, R. H. C., Thomson, R.Y., 1995: Condition fator and hepatosomatic index as estimates of energy status in male three-spined stickleback. *Journal of Fish Biology*. 47, 775-787.

Eiras, J. C., Takemoto, R. M., Pavanelli, G. C., 2006: *Métodos de estudo e técnicas laboratoriais em parasitologia de peixes*. Maringá: Eduem, Brazil.

Fortes, E., Mattos, M. J. T., Oliveira, M. F., 2000: Ocorrência de *Neoechinorhynchus curemai* (Acanthocephala) em tainha *M. curema* (Curvier et Valenciennes, 1836) do Complexo Estuário do Rio Potengí, Natal, RN, Brasil. *Revista Brasileira de Medicina Veterinária*. 22, 174-175.

Harrison, I. J., 2002: Mugilidae, Mullets. In: K.E. Carpenter (Ed). *FAO species identification guide for fishery purposes: The living marine resources of the Western Central Atlantic*. 2, 1071-1085.

Holloway, H. L., Bier, J. W., 1967: Brief notes, surveys, and comments. *Bulletin of Wildlife Disease Association*. 3, 76-77.

Huntingford, F.A., Chellappa, S., Taylor, A.C., Strang, R.H.C., 2001: Energy reserves and reproductive investment in male three-spined sticklebacks, *Gasterosteus aculeatus*. Ecology of Freshwater Fish. 10, 111-117.

Lizama, M. A. P., Takemoto, R. M., Pavanelli, G. C., 2005: Influence of host sex and age on infra communities of metazoan parasites of *Prochilodus lineatus* (Valenciennes, 1836) (Prochilodontidae) of the Upper Paraná River floodplain, Brazil. *Parasite*. 12, 299-304.

Marin, E., Baumar, J., Quintero, A., Bussière, D., Dodson, J.J., 2003: Reproduction and recruitment of white mullet (*Mugil curema*) to a tropical lagoon (Margarita Island, Venezuela) as revealed by otolith microstructure. *Fishery Bulletin*. 101, 809-821.

Martins, M. L., Fujimoto, R. Y., Andrade, P. M., Tavares-Dias, M., 2000: Recent studies on *Neoechinorhynchus curemai* Noronha, 1973 (Acanthocephala: Neoechinorhynchidae) in *Prochilodus lineatus*, Valenciennes, 1836 from Volta Grande Reservoir, MG, Brazil. *Revista Brasileira de Biologia*. 60, 673-682.

Martins, M. L., Tavares-Dias, M., Fujimoto, R. Y., Onaka, E. M., Nomura, D. T., 2004: Haematological alterations of *Leporinus macrocephalus* (Osteichthyes: Anostomidae) naturally infected by *Goezia leporini* (Nematoda: Anisakidae) in fish pond. *Arquivo Brasileiro de Medicina Veterinária e Zootecnia*. 56, 640–646.

Nickol, B. B., 1985: Epizootiology. p. 307-346. In: Crompton, D.W.T. and Nickol, B.B. (Eds). *Biology of the Acanthocephala*. Cambridge, Cambridge University Press.

Noronha, D., 1984: Remarks on *Neoechinorhynchus curemai* Noronha Eoacanthocephala, Neoechinorhynchidae). *Memórias do Instituto Oswaldo Cruz*, 79, 271.

Palm, H. W., 1997: Trypanorhynch cestodes of commercial fishes from north-east Brazilian coastal waters. *Memórias do Instituto Oswaldo Cruz*, Rio de Janeiro 92, 69-79.

Santos, R. S., Martins, M. L., Marengoni, N. G., Francisco, C. J., Piazza, R. S., Takahashi, H. K., Onaka, E. M., 2005: *Neoechinorhynchus curemai* (Acanthocephala: Neoechinorhynchidae) in *Prochilodus lineatus* (Osteichthyes: Prochilodontidae) from the Paraná River, Brazil. *Veterinary Parasitology*. 134, 111-115.

Schmidt, G. D., Roberts, L. S., 1981: *Foundations of parasitology*. St. Louis: Mosby Company.

Szpilman, M., 1992: Aqualung guide to fishes: a pratical guide to the identification of Brazilian coastal fishes. São Paulo: Aqualung Confecções, Brazil.

Sterrer, W., 1986: Phylum Acanthocephala. In: Sterrer, W., Schoepfer-Sterrer, C. (Eds). *Marine fauna and flora of Bermuda: A systematic guide to the identification of marine organisms*. 1, 223-224.

Tada, I., Otsuji, Y., Kamiya, H., Mimori, T., Sakaguchi, Y., Makizumi, S., 1983: The first case of a human infected with an acanthocephalan parasite, *Bolbosoma* sp. *Journal of Parasitology*. 69, 205-208.

Tantaleán, M., Sánchez, L., Gómez, L., Huiza, A., 2005: Acantocéfalos del Peru. *Revista Peruana de Biologia*. 12, 83-92.

Thatcher, V. E., 1991: Amazon Fish Parasites. *Amazoniana*. 11, 263-572.

Yamaguti, S., 1961: *Systema Helminthum*. v. V. *Acanthocephala*, Intersci. John Wiley and Sons Inc., New York.

Vazzoler, A. E. A. M., 1996: Biologia da reprodução de peixes teleósteos: teoria e prática. Editora da Universidade Estadual de Maringá, Brazil.

INDEX

D

E

F

G

H

I

J

L

M

N

O

P

Q

R

S

Z